BREATH

○——○

Also by James Nestor

Deep: Freediving, Renegade Science,
and What the Ocean Tells Us about Ourselves

BREATH

The New Science
of a Lost Art

James Nestor

Riverhead Books
New York
2020

RIVERHEAD BOOKS
An imprint of Penguin Random House LLC
penguinrandomhouse.com

Library of Congress Cataloging-in-Publication Data

Names: Nestor, James, author.
Title: Breath : the new science of a lost art / James Nestor.
Description: New York : Riverhead Books, 2020. | Includes bibliographical references and index.
Identifiers: LCCN 2019050863 (print) | LCCN 2019050864 (ebook) |
ISBN 9780735213616 (hardcover) | ISBN 9780735213630 (ebook)
Subjects: LCSH: Breathing exercises. | Respiration.
Classification: LCC RA782 .N47 2020 (print) | LCC RA782 (ebook) | DDC 613/.192—dc23
LC record available at https://lccn.loc.gov/2019050863
LC ebook record available at https://lccn.loc.gov/2019050864

International edition ISBN: 9780593191354

Printed in the United States of America
23rd Printing

Book design by Lauren Kolm

To K.S.

In transporting the breath, the inhalation must be full. When it is full, it has big capacity. When it has big capacity, it can be extended. When it is extended, it can penetrate downward. When it penetrates downward, it will become calmly settled. When it is calmly settled, it will be strong and firm. When it is strong and firm, it will germinate. When it germinates, it will grow. When it grows, it will retreat upward. When it retreats upward, it will reach the top of the head. The secret power of Providence moves above. The secret power of the Earth moves below.

He who follows this will live. He who acts against this will die.

—500 BCE ZHOU DYNASTY STONE INSCRIPTION

CONTENTS

Part Three — Breathing+

INTRODUCTION

The place looked like something out of Amityville: all paint-chipped walls, dusty windows, and menacing shadows cast by moonlight. I walked through a gate, up a flight of creaking steps, and knocked on the door.

When it swung open, a woman in her 30s with woolly eyebrows and oversize white teeth welcomed me inside. She asked me to take off my shoes, then led me to a cavernous living room, its ceiling painted sky blue with wispy clouds. I took a seat beside a window that rattled in the breeze and watched through jaundiced streetlight as others walked in. A guy with prisoner eyes. A stern-faced man with Jerry Lewis bangs. A blond woman with an off-center bindi on her forehead. Through the rustle of shuffling feet and whispered hellos, a truck rumbled down the street blasting "Paper Planes," the inescapable anthem of the day. I removed my belt, loosened the top button on my jeans, and settled in.

I'd come here on the recommendation of my doctor, who'd told me, "A breathing class could help." It could help strengthen my failing lungs, calm my frazzled mind, maybe give me perspective.

For the past few months, I'd been going through a rough patch. My job was stressing me out and my 130-year-old house was falling apart. I'd just recovered from pneumonia, which I'd also had the year before and the year before that. I was spending most of my time at home wheezing, working, and eating three meals a day out of the same bowl while hunched over week-old newspapers on the couch. I was in a rut—physically, mentally, and otherwise. After a few months of living this way, I took my doctor's advice and signed up for an introductory course in breathing to learn a technique called Sudarshan Kriya.

At 7:00 p.m., the bushy-browed woman locked the front door, sat in the middle of the group, inserted a cassette tape into a beat-up boom box, and pressed play. She told us to close our eyes. Through hissing static, the voice of a man with an Indian accent flowed from the speakers. It was squeaky, lilting, and too melodious to sound natural, as if it had been taken from a cartoon. The voice instructed us to inhale slowly through our noses, then to exhale slowly. To focus on our breath.

We repeated this process for a few minutes. I reached over to a pile of blankets and wrapped one around my legs to keep my stocking feet warm beneath the drafty window. I kept breathing but nothing happened. No calmness swept over me; no tension released from my tight muscles. Nothing.

Ten, maybe 20 minutes passed. I started getting annoyed and a bit resentful that I'd chosen to spend my evening inhaling dusty air on the floor of an old Victorian. I opened my eyes and looked around. Everyone had the same somber, bored look. Prisoner Eyes appeared to be sleeping. Jerry Lewis looked like he was relieving himself. Bindi sat frozen with a Cheshire Cat smile on her face. I thought about getting up and leaving, but I didn't want to be rude. The session was free; the instructor wasn't paid to be here. I needed to respect her charity. So I closed my eyes again, wrapped the blanket a little tighter, and kept breathing.

Then something happened. I wasn't conscious of any transformation taking place. I never felt myself relax or the swarm of nagging thoughts leave my head. But it was as if I'd been taken from one place and deposited somewhere else. It happened in an instant.

The tape came to an end and I opened my eyes. There was something wet on my head. I lifted my hand to wipe it off and noticed my hair was sopping. I ran my hand down my face, felt the sting of sweat in my eyes, and tasted salt. I looked down at my torso and noticed sweat blotches on my sweater and jeans. The temperature in the room was about 68 degrees—much cooler beneath the drafty window. Everyone had been covered in jackets and hoodies to keep warm. But I had somehow sweated through my clothes as if I'd just run a marathon.

The instructor approached and asked if I was OK, if I'd been sick or had a fever. I told her I felt perfectly fine. Then she said something about the body's heat, and how each inhaled breath provides us with new energy and each exhale releases old, stale energy. I tried to take it in but was having trouble focusing. I was preoccupied with how I was going to ride my bike three miles home from the Haight-Ashbury in sweat-soaked clothes.

The next day I felt even better. As advertised, there was a feeling of calm and quiet that I hadn't experienced in a long time. I slept well. The little things in life didn't bother me as much. The tension was gone from my shoulders and neck. This lasted a few days before the feeling faded out.

What exactly had happened? How did sitting cross-legged in a funky house and breathing for an hour trigger such a profound reaction?

I returned to the breathing class the next week: same experience, fewer waterworks. I didn't mention any of it to family members or friends. But I worked to understand what had happened, and I spent the next several years trying to figure it out.

Over that span of time, I fixed up my house, got out of my funk, and got a lead that might answer some of my questions about breathing. I went to Greece to write a story on freediving, the ancient practice of diving hundreds of feet below the water surface on a single breath of air. Between dives, I interviewed dozens experts, hoping to gain some perspective on what they did and why. I wanted to know how these unassuming-looking people—software engineers, advertising executives, biologists, and physicians—had trained their bodies to go without air for 12 minutes at a time, diving to depths far beyond what scientists thought possible.

When most people go underwater in a pool they bail out at ten feet after just a few seconds, ears screaming. The freedivers told me they'd previously been "most people." Their transformation was a matter of training; they'd coaxed their lungs to work harder, to tap the pulmonary capabilities that the rest of us ignore. They insisted they weren't special. Anyone in reasonable health willing to put in the hours could dive to 100, 200, even 300 feet. It didn't matter how old you were, how much you weighed, or what your genetic makeup was. To freedive, they said, all anyone had to do was master the art of breathing.

To them breathing wasn't an unconscious act; it wasn't something they just did. It was a force, a medicine, and a mechanism through which they could gain an almost superhuman power.

"There are as many ways to breathe as there are foods to eat," said one female instructor who had held her breath for more than eight minutes and once dived below 300 feet. "And each way we breathe will affect our bodies in different ways." Another diver told me that some methods of breathing will nourish our brains, while others will kill neurons; some will make us healthy, while others will hasten our death.

They told crazy stories, about how they'd breathed in ways that expanded the size of their lungs by 30 percent or more. They told me about an Indian doctor who lost several pounds by simply changing the way he inhaled, and about another man who was injected with the bacterial endotoxin *E. coli*, then breathed in a rhythmic pattern to stimulate his immune system and destroy the toxins within minutes. They told me about women who put their cancers into remission and monks who could melt circles in the snow around their bare bodies over a period of several hours. It all sounded nuts.

During my off-hours from doing underwater research, usually late at night, I read through reams of literature on the subject. Surely someone had studied the effects of this conscious breathing on landlubbers? Surely someone had corroborated the freedivers' fantastic stories of using breathing for weight loss, health, and longevity?

I found a library's worth of material. The problem was, the sources were hundreds, sometimes thousands, of years old.

Seven books of the Chinese Tao dating back to around 400 BCE focused entirely on breathing, how it could kill us or heal us, depending on how we used it. These manuscripts included detailed instructions on how to regulate the breath, slow it, hold it, and swallow it. Even earlier, Hindus considered breath and spirit the same thing, and described elaborate practices that were meant to balance breathing and preserve both physical and mental health. Then there were the Buddhists, who used breathing not only to lengthen their lives but to reach higher planes of consciousness. Breathing, for all these people, for all these cultures, was powerful medicine.

"Therefore, the scholar who nourishes his life refines the form and nourishes his breath," says an ancient Tao text. "Isn't this evident?"

Not so much. I looked for some kind of verification of these claims in more recent research in pulmonology, the medical discipline that deals with the lungs and the respiratory tract, but found next to nothing. According to what I did find, breathing technique wasn't important. Many doctors, researchers, and scientists I interviewed confirmed this position. Twenty times a minute, ten times, through the mouth, nose, or breathing tube, it's all the same. The point is to get air in and let the body do the rest.

To get a sense of how breathing is regarded by modern medical professionals, think back to your last check-up. Chances are your doctor took your blood pressure, pulse, and temperature, then placed a stethoscope to your chest to assess the health of your heart and lungs. Maybe she discussed diet, taking vitamins, stresses at work. Any issues digesting food? How about sleep? Were the seasonal allergies getting worse? Asthma? What about those headaches?

But she likely never checked your respiratory rate. She never checked the balance of oxygen and carbon dioxide in your bloodstream. How you breathe and the quality of each breath were not on the menu.

Even so, if the freedivers and the ancient texts were to be believed, how we breathe affects all things. How could it be so important and unimportant at the same time?

I kept digging, and slowly a story began to unfold. As I found out, I was not the only person who'd recently started asking these questions. While I was paging through texts and interviewing freedivers and superbreathers, scientists at Harvard, Stanford, and other renowned institutions were confirming some of the wildest stories I'd been hearing. But their work wasn't happening in the pulmonology labs. Pulmonologists, I learned, work mainly on specific maladies of the lungs—collapse, cancer, emphysema. "We're dealing with emergencies," one veteran pulmonologist told me. "That's how the system works."

No, this breathing research has been taking place elsewhere: in the muddy digs of ancient burial sites, the easy chairs of dental offices, and the rubber rooms of mental hospitals. Not the kinds of places where you'd expect to find cutting-edge research into a biological function.

Few of these scientists set out to study breathing. But, somehow, in some way, breathing kept finding them. They discovered that our capacity to breathe has changed through the long processes of human evolution, and that the way we breathe has gotten markedly worse since the dawn of the Industrial Age. They discovered that 90 percent of us—very likely me, you, and almost everyone you know—is breathing incorrectly and that this failure is either causing or aggravating a laundry list of chronic diseases.

On a more inspiring note, some of these researchers were also showing that many modern maladies—asthma, anxiety, attention deficit hyperactivity disorder, psoriasis, and more—could either be reduced or reversed simply by changing the way we inhale and exhale.

This work was upending long-held beliefs in Western medical science. Yes, breathing in different patterns really can influence our body weight and overall health. Yes, how we breathe really does affect the size and function of our lungs. Yes, breathing allows us to hack into our own nervous system, control our immune response, and restore our health. Yes, changing how we breathe will help us live longer.

No matter what we eat, how much we exercise, how resilient our genes are, how skinny or young or wise we are—none of it will matter unless we're breathing correctly. That's what these researchers discovered. The missing pillar in health is breath. It all starts there.

. . .

This book is a scientific adventure into the lost art and science of breathing. It explores the transformation that occurs inside our bodies every 3.3 seconds, the time it takes the average person to inhale and exhale. It explains how the billions and billions of molecules you bring in with each breath have built your bones, sheaths of muscle, blood, brains, and organs, and the emerging science of how these microscopic bits will influence your health and happiness tomorrow, next week, next month, next year, and decades from now.

I call this a "lost art" because so many of these new discoveries aren't new at all. Most of the techniques I'll be exploring have been around for hundreds, sometimes thousands, of years. They were created, documented, forgotten, and discovered in another culture at another time, then forgotten again. This went on for centuries.

Many early pioneers in this discipline weren't scientists. They were tinkerers, a kind of rogue group I call "pulmonauts," who stumbled on the powers of breathing because nothing else could help them. They were Civil War surgeons, French hairdressers, anarchist opera singers, Indian mystics, irritable swim coaches, stern-faced Ukrainian cardiologists, Czechoslovakian Olympians, and North Carolina choral conductors.

Few of these pulmonauts achieved much fame or respect when they were alive, and when they died their research was buried and scattered. It was even more fascinating to learn that, during the past few years, their techniques were being rediscovered and scientifically tested and proven. The fruits of this once-fringe, often forgotten research are now redefining the potential of the human body.

But why do I need to learn how to breathe? I've been breathing my whole life.

This question, which you may be asking now, has been popping up ever since I began my research. We assume, at our peril, that breathing is a passive action, just something that we do: breathe, live; stop breathing, die. But breathing is not binary. And the more I immersed myself in this subject, the more personally invested I felt about sharing this basic truth.

Like most adults, I too have suffered from a host of respiratory problems in my life. That's what landed me at the breathing class years ago. And like most people, I found that no allergy drug, inhaler, mix of supplements, or diet did much good. In the end, it was a new generation of pulmonauts who offered me a cure, and then they offered so much more.

It will take the average reader about 10,000 breaths to read from here to the end of the book. If I've done my job correctly, starting now, with every breath you take, you'll have a deeper understanding of breathing and how best to do it. Twenty times a minute, ten times, through the mouth, nose, tracheostomy, or breathing tube, it's not all the same. How we breathe really matters.

By your thousandth breath, you'll understand why modern humans are the only species with chronically crooked teeth, and why that's relevant to breathing. You'll know how our ability to breathe has deteriorated over the ages, and why our cavemen ancestors didn't snore. You'll have followed two middle-aged men as they struggle through a pioneering and masochistic 20-day study at Stanford University to test the long-held belief that the pathway through which we breathe—nose or mouth—is inconsequential. Some of what you'll learn will ruin your days and nights, especially if you snore. But in your next breaths, you'll find remedies.

By your 3,000th breath, you'll know the basics of restorative breathing. These slow and long techniques are open to everyone—old and young, sick and healthy, rich and poor. They've been practiced in Hinduism, Buddhism, Christianity, and other religions for thousands of years, but only recently have we learned how they can reduce blood pressure, boost athletic performance, and balance the nervous system.

By your 6,000th breath, you will have moved into the land of serious, conscious breathing. You'll travel past the mouth and nose, deeper into the lungs, and you'll meet a midcentury pulmonaut who healed World War II veterans of emphysema and trained Olympic sprinters to win gold medals, all by harnessing the power of the exhale.

By your 8,000th breath, you'll have pushed even deeper into the body to tap, of all things, the nervous system. You'll discover the power of overbreathing. You'll meet with pulmonauts who have used breathing to straighten scoliotic spines, blunt autoimmune diseases, and superheat themselves in subzero temperatures. None of this should be possible, and yet, as you will see, it is. Along the way, I'll be learning, too, trying to understand what happened to me in that Victorian house a decade ago.

By your 10,000th breath, and the close of this book, you and I will know how the air that enters your lungs affects every moment of your life and how to harness it to its full potential until your final breath.

This book will explore many things: evolution, medical history, biochemistry, physiology, physics, athletic endurance, and more. But mostly it will explore *you*.

By the law of averages, you will take 670 million breaths in your lifetime. Maybe you've already taken half of those. Maybe you're on breath 669,000,000. Maybe you'd like to take a few million more.

BREATH

○———○

Part One

o—o

THE EXPERIMENT

One

THE WORST BREATHERS
IN THE ANIMAL
KINGDOM

o—o

The patient arrived, pale and torpid, at 9:32 a.m. Male, middle-aged, 175 pounds. Talkative and friendly but visibly anxious. Pain: none. Fatigue: a little. Level of anxiety: moderate. Fears about progression and future symptoms: high.

Patient reported that he was raised in a modern suburban environment, bottle-fed at six months, and weaned onto jarred commercial foods. The lack of chewing associated with this soft diet stunted bone development in his dental arches and sinus cavity, leading to chronic nasal congestion.

By age 15, patient was subsisting on even softer, highly processed foods consisting mostly of white bread, sweetened fruit juices, canned vegetables, Steak-umms, Velveeta sandwiches, microwave taquitos, Hostess Sno Balls, and Reggie! bars. His mouth had become so underdeveloped it could not accommodate 32 permanent teeth; incisors and canines grew in crooked, requiring extractions, braces, retainers, and headgear to straighten. Three years of orthodontics made his small mouth even smaller, so his tongue no longer properly fit between his teeth. When he stuck it out, which he did often, visible imprints laced its sides, a precursor to snoring.

At 17, four impacted wisdom teeth were removed, which further decreased the size of his mouth while increasing his chances of developing the chronic nocturnal choking known as sleep apnea. As he aged into his 20s and 30s, his breathing became more labored and dysfunctional and his airways became more obstructed. His face would continue a vertical growth pattern that led to sagging eyes, doughy cheeks, a sloping forehead, and a protruding nose.

This atrophied, underdeveloped mouth, throat, and skull, unfortunately, belongs to me.

I'm lying on the examination chair in the Stanford Department of Otolaryngology Head and Neck Surgery Center looking at myself, looking into myself. For the past several minutes, Dr. Jayakar Nayak, a nasal and sinus surgeon, has been gingerly coaxing an endoscope camera through my nose. He's gone so deep into my head that it's come out the other side, into my throat.

"Say *eeee*," he says. Nayak has a halo of black hair, square glasses, cushioned running shoes, and a white coat. But I'm not looking at his clothes, or his face. I'm wearing a pair of video goggles that are streaming a live feed of the journey through the rolling dunes, swampy marshes, and stalactites inside my severely damaged sinuses. I'm trying not to cough or choke or gag as that endoscope squirms a little farther down.

"Say *eeee*," Nayak repeats. I say it and watch as the soft tissue around my larynx, pink and fleshy and coated in slime, opens and closes like a stop-motion Georgia O'Keeffe flower.

This isn't a pleasure cruise. Twenty-five sextillion molecules (that's

250 with 20 zeros after it) take this same voyage 18 times a minute, 25,000 times a day. I've come here to see, feel, and learn where all this air is supposed to enter our bodies. And I've come to say goodbye to my nose for the next ten days.

For the past century, the prevailing belief in Western medicine was that the nose was more or less an ancillary organ. We should breathe out of it if we can, the thinking went, but if not, no problem. That's what the mouth is for.

Many doctors, researchers, and scientists still support this position. There are 27 departments at the National Institutes of Health devoted to lungs, eyes, skin disease, ears, and so on. The nose and sinuses aren't represented in any of them.

Nayak finds this absurd. He is the chief of rhinology research at Stanford. He heads an internationally renowned laboratory focused entirely on understanding the hidden power of the nose. He's found that those dunes, stalactites, and marshes inside the human head orchestrate a multitude of functions for the body. Vital functions. "Those structures are in there for a reason!" he told me earlier. Nayak has a special reverence for the nose, which he believes is greatly misunderstood and underappreciated. Which is why he's so interested to see what happens to a body that functions without one. Which is what brought me here.

Starting today, I'll spend the next quarter of a million breaths with silicone plugs blocking my nostrils and surgical tape over the plugs to stop even the faintest amount of air from entering or exiting my nose. I'll breathe only through my mouth, a heinous experiment that will be exhausting and miserable, but has a clear point.

Forty percent of today's population suffers from chronic nasal obstruction, and around half of us are habitual mouthbreathers, with females and children suffering the most. The causes are many: dry air to stress, inflammation to allergies, pollution to pharmaceuticals. But much of the blame, I'll soon learn, can be placed on the ever-shrinking real estate in the front of the human skull.

When mouths don't grow wide enough, the roof of the mouth tends to rise up instead of out, forming what's called a V-shaped or high-arched palate. The upward growth impedes the development of the nasal cavity, shrinking it and disrupting the delicate structures in the nose. The reduced nasal space leads to obstruction and inhibits airflow. Overall, humans have the sad distinction of being the most plugged-up species on Earth.

I should know. Before probing my nasal cavities, Nayak took an X-ray of my head, which provided a deli-slicer view of every nook and cranny in my mouth, sinuses, and upper airways.

"You've got some . . . *stuff*," he said. Not only did I have a V-shaped palate, I also had "severe" obstruction to the left nostril caused by a "severely" deviated septum. My sinuses were also riddled with a profusion of deformities called *concha bullosa*. "Super uncommon," said Nayak. It was a phrase nobody wants to hear from a doctor.

My airways were such a mess that Nayak was amazed I hadn't suffered from even more of the infections and respiration problems I'd known as a kid. But he was reasonably certain I could expect some degree of serious breathing problems in the future.

Over the next ten days of forced mouthbreathing, I'll be putting myself inside a kind of mucousy crystal ball, amplifying and hastening the deleterious effects on my breathing and my health, which will keep getting worse as I get older. I'll be lulling my body into a state it already knows, that half the population knows, only multiplying it many times.

"OK, hold steady," Nayak says. He grabs a steel needle with a wire brush at the end, about the size of a mascara brush. I'm thinking, *He's not going to put that thing up my nose.* A few seconds later, he puts that thing up my nose.

I watch through the video goggles as Nayak maneuvers the brush deeper. He keeps sliding until it is no longer up my nose, no longer playing around my nasal hair, but wiggling inside of my head a few inches deep. "Steady, steady," he says.

When the nasal cavity gets congested, airflow decreases and bacteria flourish. These bacteria replicate and can lead to infections and colds and more congestion. Congestion begets congestion, which gives us no other option but to habitually breathe from the mouth. Nobody knows how soon this damage occurs. Nobody knows how quickly bacteria accumulate in an obstructed nasal cavity. Nayak needs to grab a culture of my deep nasal tissue to find out.

I wince as I watch him twist the brush deeper still, then spin it, skimming off a layer of gunk. The nerves this far up the nose are designed to feel the subtle flow of air and slight modulations in air temperature, not steel brushes. Even though he's dabbed an anesthetic in there, I can still feel it. My brain has a hard time knowing exactly what to do, how to react. It's difficult to explain, but it feels like someone is needling a conjoined twin that exists somewhere outside of my own head.

"The things you never thought you'd be doing with your life," Nayak laughs, putting the bleeding tip of the brush into a test tube. He'll compare the 200,000 cells from my sinuses with another sample ten days from now to see how nasal obstruction affects bacterial growth. He shakes the test tube, hands it to his assistant, and politely asks me to take the video goggles off and make room for his next patient.

Patient #2 is leaning against the window and snapping photos with his phone. He's 49 years old, deeply tanned with white hair and Smurf-blue eyes, and he's wearing spotless beige jeans and leather loafers without socks. His name is Anders Olsson, and he's flown 5,000 miles from Stockholm, Sweden. Along with me, he's ponied up more than $5,000 to join the experiment.

I'd interviewed Olsson several months ago after coming across his website. It had all the red flags of flakiness: stock images of blond women striking hero poses on mountaintops, neon colors, frantic use of exclamation points, and bubble fonts. But Olsson wasn't some fringe character. He'd spent ten years collecting and conducting serious scientific research. He'd written dozens of posts and self-published a book explaining breathing from the subatomic level on up, all annotated with hundreds of studies. He'd also become one of Scandinavia's most respected and popular breathing therapists, helping to heal thousands of patients through the subtle power of healthy breathing.

When I mentioned during one of our Skype conversations that I would be mouthbreathing for ten days during an experiment, he cringed. When I asked if he wanted to join in, he refused. "I do not want to," he declared. "But I am curious."

Now, months later, Olsson plops his jet-lagged body onto the examination chair, puts on the video glasses, and inhales one of his last nasal breaths for the next 240 hours. Beside him, Nayak twirls the steel endoscope the way a heavy metal drummer handles a drumstick. "OK, lean your head back," says Nayak. A twist of the wrist, a crane of the neck, and he goes deep.

The experiment is set up in two phases. Phase I consists of plugging our noses and attempting to live our everyday lives. We'll eat, exercise, and sleep as usual, only we'll do it while breathing only through our mouths. In Phase II, we'll eat, drink, exercise, and sleep like we did during Phase I, but we'll switch the pathway and breathe through our noses and practice a number of breathing techniques throughout the day.

Between phases we'll return to Stanford and repeat all the tests we've just taken: blood gases, inflammatory markers, hormone levels, smell, rhinometry, pulmonary function, and more. Nayak will compare data sets and see what, if anything, changed in our brains and bodies as we shifted our style of breathing.

I'd gotten a fair share of gasps from friends when I told them about the experiment. "Don't do it!" a few yoga devotees warned. But most people just shrugged. "I haven't breathed out of my nose in a decade," said a friend who had suffered allergies most of his life. Everyone else said the equivalent of: *What's the big deal? Breathing is breathing.*

Is it? Olsson and I will spend the next 20 days finding out.

. . .

A while back, some 4 billion years ago, our earliest ancestors appeared on some rocks. We were small then, a microscopic ball of sludge. And we were hungry. We needed energy to live and proliferate. So we found a way to eat air.

The atmosphere was mostly carbon dioxide then, not the best fuel, but it worked well enough. These early versions of us learned to take this gas in, break it down, and spit out what was left: oxygen. For the next billion years, the primordial goo kept doing this, eating more gas, making more sludge, and excreting more oxygen.

Then, around two and a half billion years ago, there was enough oxygen waste in the atmosphere that a scavenger ancestor emerged to make use of it. It learned to gulp in all that leftover oxygen and excrete carbon dioxide: the first cycle of aerobic life.

Oxygen, it turned out, produced 16 times more energy than carbon dioxide. Aerobic life forms used this boost to evolve, to leave the sludge-covered rocks behind and grow larger and more complex. They crawled up to land, dove deep into the sea, and flew into the air. They became plants, trees, birds, bees, and the earliest mammals.

Mammals grew noses to warm and purify the air, throats to guide air into lungs, and a network of sacs that would remove oxygen from the atmosphere and transfer it into the blood. The aerobic cells that once clung to swampy rocks so many eons ago now made up the tissues in mammalian bodies. These cells took oxygen from our blood and returned carbon dioxide, which traveled back through the veins, through the lungs, and into the atmosphere: the process of breathing.

The ability to breathe so efficiently in a wide variety of ways—consciously and unconsciously; fast, slow, and not at all—allowed our mammal ancestors to catch prey, escape predators, and adapt to different environments.

It was all going so well until about 1.5 million years ago, when the pathways through which we took in and exhaled air began to shift and fissure. It was a shift that, much later in history, would affect the breathing of every person on Earth.

I'd been feeling these cracks for much of my life, and chances are you have, too: stuffy noses, snoring, some degree of wheezing, asthma, allergies, and the rest. I'd always thought they were a normal part of being human. Nearly everyone I knew suffered from one problem or another.

But I came to learn that these problems didn't randomly develop. Something caused them. And the answers could be found in a common and homely human trait.

A few months before the Stanford experiment, I flew to Philadelphia to visit Dr. Marianna Evans, an orthodontist and dental researcher who'd spent the last several years looking into the mouths of human skulls, both ancient and modern. We were standing in the basement of the University of Pennsylvania Museum of Archaeology and Anthropology, surrounded by several hundred specimens. Each was engraved with letters and numbers and stamped with its "race": *Bedouin, Copt, Arab of Egypt, Negro Born in Africa.* There were Brazilian prostitutes, Arab slaves, and Persian prisoners. The most famous specimen, I was told, came from an Irish prisoner who'd been hanged in 1824 for killing and eating fellow convicts.

The skulls ranged from 200 to thousands of years old. They were part of the Morton Collection, named after a racist scientist named Samuel Morton, who, starting in the 1830s, collected skeletons in a failed attempt to prove the superiority of the Caucasian race. The only positive outcome of Morton's work is that the skulls he spent so many years gathering now provide a snapshot of how people used to look and breathe.

Where Morton claimed to see inferior races and genetic "degradation," Evans discovered something close to perfection. To demonstrate what she meant, she walked over to a cabinet and retrieved a skull marked *Parsee,* for Persian, from behind the protective glass. She wiped bone dust on the sleeve of her cashmere sweater and ran a neatly trimmed fingernail along its jaw and face.

"These are twice as large as they are today," she said in a staccato Ukrainian accent. She was pointing at the nasal apertures, the two holes that connect the sinuses to the back of the throat. She turned the skull around so it was staring at us. "So wide and pronounced," she said approvingly.

Evans and her colleague Dr. Kevin Boyd, a Chicago-based pediatric dentist, have spent the last four years X-raying more than 100 skulls from the Morton Collection and measuring the angles from the ear hole to the nose and from the forehead to the chin. These measurements, which are called the Frankfort plane and N-perpendicular, show the symmetry of each specimen, how well proportioned the mouth was relative to the face, the nose to the palate, and, to a large extent, how well the people who owned these skulls might have breathed.

Every one of the ancient skulls was identical to the *Parsee* sample. They all had enormous forward-facing jaws. They had expansive sinus cavities and broad mouths. And, bizarrely, even though none of the ancient people ever flossed, or brushed, or saw a dentist, they all had straight teeth.

The forward facial growth and large mouths also created wider airways. These people very likely never snored or had sleep apnea or sinusitis or many other chronic respiratory problems that affect modern populations. They did not because they could not. Their mouths were far too large, and their airways too wide for anything to block them. They breathed easy. Nearly all ancient humans shared this forward structure—not just in the Morton Collection, but everywhere around the world. This remained true from the time when *Homo sapiens* first appeared, some 300,000 years ago, to just a few hundred years ago.

Evans and Boyd then compared the ancient skulls to the modern skulls of their own patients and others. Every modern skull had the opposite growth pattern, meaning the angles of the Frankfort plane and N-perpendicular were reversed: chins had recessed behind foreheads, jaws were slumped back, sinuses shrunken. All the modern skulls showed some degree of crooked teeth.

Of the 5,400 different species of mammals on the planet, humans are now the only ones to routinely have misaligned jaws, overbites, underbites, and snaggled teeth, a condition formally called malocclusion.

To Evans, this raised a fundamental question: "Why would we evolve to make ourselves sick?" she asked. She put the *Parsee* skull back in the cabinet and took out another labeled *Saccard*. Its perfect facial form was a mirror image of the others. "That's what we're trying to find out," she said.

Evolution doesn't always mean progress, Evans told me. It means change. And life can change for better or worse. Today, the human body is changing in ways that have nothing to do with the "survival of the fittest." Instead, we're adopting and passing down traits that are detrimental to our health. This concept, called *dysevolution*, was made popular by Harvard biologist Daniel Lieberman, and it explains why our backs ache, feet hurt, and bones are growing more brittle. Dysevolution also helps explain why we're breathing so poorly.

To understand how this all happened, and why, Evans told me, we need to go back in time. Way back. To before *Homo sapiens* were even *sapiens*.

· · ·

What strange creatures. Standing in the tall grass of the savanna, all gangly arms and pointy elbows, gazing out into the wide, wild world from foreheads that looked like hairy visors. As the breeze swayed the grass, our nostrils, the size of gum drops, flexed vertically above our chinless mouths, picking up whatever scents the wind brought in.

The time was 1.7 million years ago, and the first human ancestor, *Homo habilis*, was roaming the eastern shores of Africa. We'd long since left the trees, learned to walk on our legs, and trained ourselves to use the small "finger" on the inside of our hands, to turn it upside down into an opposable thumb. We used this thumb and fingers to grab things, to pull plants and roots and grasses from the ground, and to build hunting tools from stone that were sharp enough to carve tongues out of antelope and strip meat from bone.

Eating this raw diet took a lot of time and effort. So we gathered stones and bashed prey against rocks. Tenderizing food, especially meat, spared us from some of the effort of digesting and chewing, which saved energy. We used this extra energy to grow a larger brain.

Grilling food was even better. Around 800,000 years ago, we began processing food in fire, which released an enormous amount of additional calories. Our large intestines, which helped break down rough and fibrous fruits and vegetables, would shrink considerably under this new diet, and that change alone saved even more energy. These more modern ancestors, *Homo erectus,* used it to grow an even bigger brain—an astounding 50 percent larger than those of our *habilis* ancestors.

We began to look less like apes and more like people. If you could take a *Homo erectus,* dress him in a Brooks Brothers suit, and put him on a subway, he probably wouldn't draw a second glance. These ancient ancestors were so similar to us that they very likely could have had our children.

The innovation of mashing and cooking food, however, had consequences. The quickly growing brain needed space to stretch out, and it took it from the front of our faces, home to sinuses, mouths, and airways. Over time, muscles at the center of the face loosened, and bones in the jaw weakened and grew thinner. The face shortened and the mouth shrank, leaving behind a bony protuberance that replaced the squashed snout of our ancestors. This new feature was ours alone and distinguished us from other primates: the protruding nose.

The problem was that this smaller, vertically positioned nose was less efficient at filtering air, and it exposed us to more airborne pathogens and bacteria. The smaller sinuses and mouth also reduced space in our throats. The more we cooked, the more soft, calorie-rich food we consumed, the larger our brains grew and the tighter our airways became.

Homo sapiens first emerged on the African savanna around 300,000 years ago. We were among a coterie of other human species: *Homo heidelbergensis,* a robust creature who built shelters and hunted big game in what is now Europe; *Homo neanderthalensis* (Neanderthals), with their massive noses and stunted limbs, who learned to make clothes and flourish in frigid environments; and *Homo naledi,* a throwback to early ancestors, with tiny brains, flared hips, and spindly arms that hung down from squat bodies.

What a sight it might have been, these ragtag species all gathered around a blazing campfire at night, a Star Wars cantina of early humanity, sipping river water from palm cups, picking grubs from each other's hair, comparing the ridges of their brows, and scampering off behind boulders to have interspecies sex in the glow of starlight.

Then, no more. The big-nosed Neanderthals, the scrawny *naledi*, the thick-necked *heidelbergensis* were all killed off by disease, weather, each other, or animals, or laziness, or something else. There was only one human left in the long family tree: us.

In colder climates, our noses would grow narrower and longer to more efficiently heat up air before it entered our lungs; our skin would grow lighter to take in more sunshine for production of vitamin D. In sunny and warm environments, we adapted wider and flatter noses, which were more efficient at inhaling hot and humid air; our skin would grow darker to protect us from the sun. Along the way, the larynx would descend in the throat to accommodate another adaptation: vocal communication.

The larynx works as a valve to shuttle food into the stomach and protect us from inhaling it and other objects. Every animal, and every other *Homo* species, had evolved a higher larynx, located toward the top of the throat. This made sense, since a high larynx functions most efficiently, allowing the body to rid itself quickly should anything get stuck in our airways.

As humans developed speech, the larynx sank, opening up space in the back of the mouth and allowing a wider range of vocalizations and volumes. Smaller lips were easier to manipulate, and ours evolved to be thinner and less bulbous. More nimble and flexible tongues made it easier to control the nuance and structure of sounds, so the tongue slipped farther down the throat and pushed the jaw forward.

But this lowered larynx became less efficient at its original purpose. It created too much space at the back of the mouth and made early humans susceptible to choking. We could choke if we swallowed something too big, and we'd choke on smaller objects that were swallowed quickly and sloppily. *Sapiens* would become the only animals, and the only human species, that could easily choke on food and die.

Strangely, sadly, the same adaptations that would allow our ancestors to outwit, outmaneuver, and outlive other animals—a mastery of fire and processing food, an enormous brain, and the ability to communicate in a vast range of sounds—would obstruct our mouths and throats and make it much harder for us to breathe. This recessed growth would, much later, make us prone to choke on our own bodies when we slept: to snore.*

None of this mattered to the early humans, of course. For tens of thousands of years, our ancestors would use their wildly developed heads to breathe just fine. Armed with a nose, a voice, and a supersized brain, humans took over the world.

. . .

I'd been thinking about our hirsute forebears ever since I'd visited Evans months back. There they were, crouched along the rocky African shore, articulating the first vowels with their flexible lips, pulling in easy breaths through gaping nasal apertures, and chomping on braised rabbit with perfect teeth.

And here I am, slack-jawed under an LED light, staring at Wikipedia's *Homo floresiensis* page on my phone, chewing on bits of a low-carb

---

*Pugs, mastiffs, boxers, and other brachycephalic dogs have been bred to have flat faces and smaller sinus cavities, and, as such, suffer from a similar range of chronic respiratory problems. In many ways, modern humans have become the *Homo* equivalent of these highly inbred dogs.

nutritional bar with crooked teeth, coughing and wheezing and sucking exactly no air through my obstructed nose.

It's evening on the second day of the Stanford mouthbreathing experiment, and I'm in bed with silicone plugs jammed inside my nasal cavities, covered with tape. For the past few nights I've been splayed out in a part of my house usually reserved for relatives and friends. I had a feeling that my mouthbreathing lifestyle might be a challenge for my wife. Lying here, tossing and turning, thinking about cavemen and unable to sleep, I'm happy I moved.

I've got a pulse oximeter device about the size of a matchbook strapped to my wrist. There's a glowing red wire extending from it and wrapping around my middle finger. Every few seconds, the device records my heart rate and blood oxygen levels, using this information to assess how often and how severely my too-deep tongue might get lodged in my too-small mouth and cause me to hold my breath, a condition more commonly known as sleep apnea.

To gauge the severity of my snoring and apnea, I've downloaded a phone app that records a constant stream of audio through the night, then provides a minute-to-minute graph of my breathing health every morning. A night vision security camera just above the bed monitors every movement.

Inflammation in the throat and polyps both contribute to snoring and sleep apnea. Nasal obstruction triggers this nighttime choking as well, but nobody knows how quickly the damage comes on, or how severe it might become. Before now, nobody had tested it.

Last night, in my first run of self-inflicted nasal obstructed sleeping, my snoring increased by 1,300 percent, to 75 minutes through the night. Olsson's numbers were even worse. He went from zero to four hours, ten minutes. I'd also suffered a fourfold increase in sleep apnea events. All this, in just 24 hours.

Now, lying here again, no matter how I try to relax and submit to this experiment, it's a challenge. Every 3.3 seconds another blast of unfiltered, unmoistened, and unheated air enters through my mouth—drying my tongue, irritating my throat, and pissing off my lungs. And I've got 175,000 more breaths to go.

Two

MOUTHBREATHING

○——○

It's 8:15 a.m., and Olsson bursts in, Kramer-style, through the side door of the downstairs flat I'm in. "Good morning," he shouts. He has little balls of silicone lodged up his nose and is wearing cut-off sweatpants and an Abercrombie & Fitch sweatshirt.

Olsson rented a studio apartment across the street from me for the month, close enough to sneak over wearing his pajamas, but not close enough to avoid looking like a freak doing it. His face, once tan and bright, is now gaunt and sallow, and he looks like Nick Nolte in that police mugshot. He has the same spaced-out expression that he had yesterday; the same haunted grin that he wore the day before and the day before that.

Today marks the halfway point of the mouthbreathing phase of the experiment. And today, like every other day, as he's been doing three times a day—morning, noon, and night—Olsson takes a seat across from me at the table. *One-two-three*, we flip on a pile of beeping and burping machines lumped on the table, strap cuffs to our arms, place sensors on our ears, stick thermometers in our mouths, and begin recording our physiological data on spreadsheets. The readouts reveal what the previous days have revealed: mouthbreathing is destroying our health.

My blood pressure has spiked by an average of 13 points from where it was before the test, which puts me deep into stage 1 hypertension. If left unchecked, this state of chronically raised blood pressure, also shared by a third of the U.S. population, can cause heart attacks, stroke, and other serious problems. Meanwhile, my heart rate variability, a measure of nervous system balance, has plummeted, suggesting that my body is in a state of stress. Then there's my pulse, which has increased, and my body temperature, which has decreased, and my mental clarity, which has hit rock bottom. Olsson's data mirror mine.

But the worst part about all this is how we *feel*: awful. Every day it all just seems to be getting worse. And every day, at this exact time, Olsson finishes off his last test, removes the respirator mask from his cotton-white hair, stands up, and jams the silicone plugs a little deeper into his nostrils. He puts his sweatshirt back on and says, "I'll see you later," then walks out the door. I nod and watch as he trots his slippered feet through the hallways and back across the street.

The final testing protocol, eating, happens alone. Through both phases of the experiment, we'll be eating the same food at the same time and continuously recording our blood sugar levels while taking the same amount of steps throughout the day to see how mouthbreathing and nasal breathing might affect weight and metabolism. Today it's three eggs, half an avocado, a piece of German brown bread, and a pot of Lapsang tea. Which means that, ten days from now, I will again be sitting in this kitchen, eating this same meal.

After eating, I do the dishes, clean up used filters, pH strips, and

Post-it notes in the living room laboratory, and answer some emails. Sometimes Olsson and I sit around and experiment with more comfortable and effective ways to keep our noses blocked: waterproof earplugs (too hard), foam earplugs (too soft), a swimmer's nose clip (too painful), a CPAP nose pillow (comfortable, but it looks like a bondage device), toilet paper (too airy), chewing gum (too slimy), and, finally, surgical tape over silicone or foam earplugs, which is chafing and stifling but the least atrocious of the options.

But most of the time, all day, every day, for the past five days, Olsson and I have just sat around alone in our apartments and hated life. I often feel as though I'm trapped in some sad sitcom in which nobody laughs, a Groundhog Day of perpetual and unending misery.

Luckily, today is a little different. Today, Olsson and I are going on a bike ride. Not on a beach boardwalk or in the shadow of the Golden Gate, but inside the concrete walls of a fluorescent-lit neighborhood gym.

The cycling was Olsson's idea. He'd spent about ten years researching the differences in performance between nasal breathers and mouthbreathers during intense exercise. He'd conducted his own studies on CrossFit athletes, and he'd worked with coaches. He'd become convinced that mouthbreathing can put the body into a state of stress that can make us more quickly fatigued and sap athletic performance. He insisted that, for a few days during each phase of the experiment, we saddle up on stationary bikes and pedal to the edge of our aerobic capacity. The plan was to meet at the gym at 10:15 a.m.

I put on some shorts, grab the fitness tracker, an extra set of silicone plugs, a water bottle, and exit through the backyard. Waiting by the fence is Antonio, a contractor and longtime friend who has been doing renovation work on an upper floor of my house. He looks over, and before I can make a beeline for the garden exit, he notices the pink earplugs in my nose, drops an armful of two-by-fours, and comes over to take a closer look.

I'd known Antonio for 15 years and he'd heard about the oddball stories in far-off places I'd researched in the past. He'd always been interested and supportive. That ended when I tell him about what I've been up to this week.

"This is a bad idea," he says. "In school, when I was young, teachers walked around the classroom, man, and *pop-pop-pop*." He smacks the back of his own head for emphasis. "You're breathing from your mouth, you get *pop*," he says. Mouthbreathing leads to sickness and is disrespectful, he told me, which is why he and everyone else he grew up with in Puebla, Mexico, learned to breathe through the nose.

Antonio told me his partner, Janet, suffers from chronic obstruction and runny nose. Janet's son, Anthony, is also a chronic mouthbreather. He's starting to suffer the same problems. "I keep telling them this is bad, they try to fix it," Antonio said. "But it's hard, man."

I'd heard a similar story from an Indian-British man named David a few days ago, when Olsson and I attempted our first nasal-obstructed jog along the Golden Gate Bridge. David noticed our nasal bandages, stopped us, and asked what we were doing. Then he told us how he'd had obstruction problems all his life. "Always plugged or running, it never seemed to be, you know, open," he said. He'd spent the last 20 years squirting various drugs up his nostrils, but they became less effective over time. Now he'd developed chronic respiratory problems.

To avoid hearing more of these stories and to evade any more unwanted attention, I'd learned to go outside only when I had to. Don't get me wrong: San Franciscans love weirdos. There was once a guy who used to walk Haight Street with a hole in the back of his jeans so that his tail—an *actual* human tail about five inches long—could swing freely behind him. He hardly got second glances.

But the sight of Olsson and me with plugs and tape and whatever else in and around our noses has proven too much for locals to bear. Everywhere we go, we get either questioned or somebody's long life story of breathing woes, how he is congested, how her allergies keep getting worse, how his head hurts and sleep suffers the worse his breathing seems to get.

I wave goodbye to Antonio, pull the visor of the baseball cap a little farther down to hide my plugged face and jog a few blocks to the gym. I make my way around women speedwalking on treadmills and old men on weight machines. I can't help noticing that all of them are mouthbreathing.

Then I boot up the pulse oximeter, set the stop watch, hop on a stationary bike, latch my feet into the pedals, and I'm off.

The bike experiment is a repeat of several studies conducted 20 years earlier by Dr. John Douillard, a trainer to elite athletes, from tennis star Billie Jean King to triathletes to the New Jersey Nets. In the 1990s, Douillard became convinced that mouthbreathing was hurting his clients. To prove it, he gathered a group of professional cyclists, rigged them up with sensors to record their heart rate and breathing rate, and put them on stationary bikes. Over several minutes, Douillard increased the resistance on the pedals, requiring the athletes to exert progressively more energy as the experiment went on.

During the first trial, Douillard told the athletes to breathe entirely through their mouths. As the intensity increased, so did the rate of breathing, which was expected. By the time athletes reached the hardest stage of the test, pedaling out 200 watts of power, they were panting and struggling to catch a breath.

Then Douillard repeated the test while the athletes breathed through their noses. As the intensity of exercise increased during this phase, the rate of breathing *decreased*. At the final, 200-watt stage, one subject who had been mouthbreathing at a rate of 47 breaths per minute was nasal breathing at a rate of 14 breaths a minute. He maintained the same heart rate at which he'd started the test, even though the intensity of the exercise had increased tenfold.

Simply training yourself to breathe through your nose, Douillard reported, could cut total exertion in half and offer huge gains in endurance. The athletes felt invigorated while nasal breathing rather than exhausted. They all swore off breathing through their mouths ever again.

For the next 30 minutes on the stationary bike, I'll follow Douillard's test protocol, but instead of measuring exertion with weight, I'll use distance. I'll keep my heart rate locked in to a consistent 136 beats per minute while measuring how far I can go with my nose plugged and breathing only from my mouth. Olsson and I will come back here over the next several days, then return next week to repeat the test while breathing only through our noses. This data will provide a general overview of how these two breathing channels affect endurance and energy efficiency.

To understand how and why Douillard's experiment worked, we first need to understand the ways the body makes energy from air and food. There are two options: with oxygen, a process known as aerobic respiration, and without it, which is called anaerobic respiration.

Anaerobic energy is generated only with glucose (a simple sugar), and it's quicker and easier for our bodies to access. It's a kind of backup system and turbo boost when the body doesn't have enough oxygen. But anaerobic energy is inefficient and can be toxic, creating an excess of lactic acid. The nausea, muscle weakness, and sweating you experience after you've pushed it too hard at the gym is the feeling of anaerobic overload. This process explains why the first few minutes of an intense workout are often so miserable. Our lungs and respiratory system haven't caught up to supply the oxygen our bodies need, and so the body has to use anaerobic respiration. This also explains why, after we're warmed up, exercise feels easier. The body has switched from anaerobic to aerobic respiration.

These two energies are made in different muscle fibers throughout the body. Because anaerobic respiration is intended as a backup system, our bodies are built with fewer anaerobic muscle fibers. If we rely on these less-developed muscles too often, they eventually break down. More injuries occur during the post–New Year's rush to gyms than at any other time of the year, because too many people attempt to exercise far over their thresholds. Essentially, anaerobic energy is like a muscle car—it's fast and responsive for quick trips, but polluting and impractical for long hauls.

This is why aerobic respiration is so important. Remember those cells that evolved to eat oxygen 2.5 billion years ago and kicked off an explosion of life? We've got some 37 trillion of them in our bodies. When we run our cells aerobically with oxygen, we gain some 16 times more energy efficiency over anaerobic. The key for exercise, and for the rest of life, is to stay in that energy-efficient, clean-burning, oxygen-eating aerobic zone for the vast majority of time during exercise and at all times during rest.

Back in the gym, I pedal a little harder, breathe a little deeper, and watch as my heart rate increases steadily, from 112 to 114 and on up. Over the next three minutes of warm-up, I need to get to 136 then keep it there for a half hour. This rate should be right at the aerobic/anaerobic threshold for a man my age.

In the 1970s, Phil Maffetone, a top fitness coach who worked with Olympians, ultramarathoners, and triathletes, discovered that most standardized workouts could be more injurious than beneficial to athletes. The reason is that everybody is different, and everybody will react differently to training. Busting out a hundred pushups may be great for one person but harmful to another. Maffetone personalized his training to focus on the more subjective metric of heart rates, which ensured that his athletes stayed inside a defined aerobic zone, and that they burned more fat, recovered faster, and came back the next day—and the next year—to do it again.

Finding the best heart rate for exercise is easy: subtract your age from 180. The result is the maximum your body can withstand to stay in the aerobic state. Long bouts of training and exercise can happen below this rate but never above it, otherwise the body will risk going too deep into the anaerobic zone for too long. Instead of feeling invigorated and strong after a workout, you'd feel tired, shaky, and nauseated.

Which is basically what happens to me. After I do a half hour of vigorous pedaling and openmouthed huffing, the clock on the stationary bike ticks down to zero and the whirling gears slow to a stop. I'm sweating profusely and feel bleary-eyed, but I've pedaled a total of only 6.44 miles. I scoot off the bike and let Olsson take a spin, then it's back to the home lab for a shower, a glass of water, and more testing.

.   .   .

Decades before Olsson and I jammed our noses shut, and before Douillard put his cyclists through the rounds, scientists were running their own tests on the pros and cons of mouthbreathing.

There was Austen Young, an enterprising doctor in England who, in the 1960s, treated a slew of chronic nose-bleeders by sewing their nostrils shut. One of Young's followers, Valerie J. Lund, revived the procedure in the 1990s and stitched the nostrils of dozens of patients. I repeatedly tried to contact Lund to ask how her mouthbreathing patients fared after weeks, months, and years, but never got a reply. Luckily, those consequences were spelled out by a Norwegian-American orthodontist and researcher chasing very different ends.

Egil P. Harvold's hideous experiments in the 1970s and 80s would not go over well with PETA or with anyone who has ever really cared for animals. Working from a lab in San Francisco, he gathered a troop of rhesus monkeys and stuffed silicone deep into the nasal cavities of half of them, leaving the other half as they were. The obstructed animals couldn't remove the plugs, and they couldn't breathe at all through their noses. They were forced to adapt to constant mouthbreathing.

Over the next six months, Harvold measured the animals' dental arches, the angles of their chins, the length of their faces, and more. The plugged-up monkeys developed the same downward growth pattern, the same narrowing of the dental arch, and gaping mouth. Harvold repeated these experiments, this time keeping animals obstructed for two years. They fared even worse. Along the way, he took a lot of pictures.

The photographs are heart-wrenching, not only for the sake of the poor monkeys, but because they also offer such a clear reflection of what happens to our own species: after just a few months, faces grew long, slack-jawed, and glazed over.

Mouthbreathing, it turns out, changes the physical body and transforms airways, all for the worse. Inhaling air through the mouth decreases pressure, which causes the soft tissues in the back of the mouth to become loose and flex inward, creating less space and making breathing more difficult. Mouthbreathing begets more mouthbreathing.

Inhaling from the nose has the opposite effect. It forces air against all those flabby tissues at the back of the throat, making the airways wider and breathing easier. After a while, these tissues and muscles get "toned" to stay in this opened and wide position. Nasal breathing begets more nasal breathing.

"Whatever happens to the nose affects what's happening in the mouth, the airways, the lungs," said Patrick McKeown during a phone interview. He's a bestselling Irish author and one of the world's leading experts on nasal breathing. "These aren't separate things that operate autonomously—it's one united airway," he told me.

None of this should come as a surprise. When seasonal allergies hit, incidences of sleep apnea and breathing difficulties shoot up. The nose gets stuffed, we start mouthbreathing, and the airways collapse. "It's simple physics," McKeown told me.

Sleeping with an open mouth exacerbates these problems. Whenever we put our heads on a pillow, gravity pulls the soft tissues in the throat and tongue down, closing off the airway even more. After a while, our airways get conditioned to this position; snoring and sleep apnea become the new normal.

.   .   .

It's the last night of the nasal obstruction phase of the experiment, and I am, again, sitting up in bed and staring out the window.

When a Pacific breeze blows in, which it does most nights, the shadows of the trees and plants on the backyard wall across from my bedroom start to move and groove in a chromatic kaleidoscope. One moment they reorganize into a cadre of Edward Gorey gentlemen in waistcoats, the next into crooked Escher staircases. Another gust of wind, and these scenes disintegrate and reform into recognizable stuff: ferns, bamboo leaves, bougainvillea.

This is a long way of saying: I can't sleep. My head's been propped up on pillows and I've been taking notes on this creepy tableau for 15, 20, maybe 40 minutes. I unconsciously try to sniff and clear my nose, but instead get a jolt of pain in my head. It's a sinus headache, and in my case, self-inflicted.

Each night for the past week and a half, I've felt as if I was getting softly choked to death in my sleep and my throat was closing in on itself. Because it is, and because I am. Forced mouthbreathing was very likely changing the shape of my airways, just as it did with Harvold's monkeys.

The changes weren't happening in a matter of months, either, but days. It was getting worse with every breath I took.

My snoring has increased 4,820 percent from ten days ago. For the first time that I'm aware of, I'm beginning to suffer from obstructive sleep apnea. At my worst, I've averaged 25 "apnea events," meaning I was choking so severely that my oxygen levels dropped to 90 percent or below.

Whenever oxygen falls below 90 percent, the blood can't carry enough of it to support body tissues. If this goes on too long, it can lead to heart failure, depression, memory problems, and early death. My snoring and sleep apnea are still far below that of any medically diagnosed condition, but these scores were getting worse the longer I stayed plugged up.

Every morning Olsson and I would listen to recordings of ourselves sleeping the night before. We laughed at first, then we got a bit frightened: what we heard weren't the sounds of happy Dickensian drunks, but of men being strangled to death by our own bodies.

"More wholesome to sleep . . . with the mouth shut," wrote Levinus Lemnius, a Dutch physician from the 1500s who was credited as one of the first researchers to study snoring. Even back then, Levinus knew how injurious obstructive breathing during sleep could be. "For they that sleep with their Jaws extended, by reason of their breath, and the air tossed to and fro, have their tongues and palates dry, and desire to be moistened by drinking in the night."

This was another thing that kept happening to me. Mouthbreathing causes the body to lose 40 percent more water. I felt this all night, every night, waking up constantly parched and dry. You'd think this moisture loss would decrease the need to urinate, but, oddly, the opposite was true.

During the deepest, most restful stages of sleep, the pituitary gland, a pea-size ball at the base of the brain, secretes hormones that control the release of adrenaline, endorphins, growth hormone, and other substances, including vasopressin, which communicates with cells to store more water. This is how animals can sleep through the night without feeling thirsty or needing to relieve themselves.

But if the body has inadequate time in deep sleep, as it does when it experiences chronic sleep apnea, vasopressin won't be secreted normally. The kidneys will release water, which triggers the need to urinate and signals to our brains that we should consume more liquid. We get thirsty, and we need to pee more. A lack of vasopressin explains not only my own irritable bladder but the constant, seemingly unquenchable thirst I have every night.

There are several books that describe the horrendous health effects of snoring and sleep apnea. They explain how these afflictions lead to bed-wetting, attention deficit hyperactivity disorder (ADHD), diabetes, high blood pressure, cancer, and so on. I'd read a report from the Mayo Clinic which found that chronic insomnia, long assumed to be a psychological problem, is often a breathing problem. The millions of Americans who have a chronic insomnia disorder and who are, right now, like me, staring out bedroom windows, or at TVs, phones, or ceilings, can't sleep because they can't breathe.

And contrary to what most of us might think, no amount of snoring is normal, and no amount of sleep apnea comes without risks of serious health effects. Dr. Christian Guilleminault, a sleep researcher at Stanford, found that children who experienced no apnea events at all—only heavy breathing and light snoring, or "increased respiratory effort"—could suffer from mood disorders, blood pressure derangements, learning disabilities, and more.

Mouthbreathing was also making me dumber. A recent Japanese study showed that rats who had their nostrils obstructed and were forced to breathe through their mouths developed fewer brain cells and took twice as long to make their way through a maze than nasal-breathing controls. Another Japanese study in humans from 2013 found that mouthbreathing delivered a disturbance of oxygen to the prefrontal cortex, the area of the brain associated with ADHD. Nasal breathing had no such effects.

The ancient Chinese were onto it as well. "The breath inhaled through the mouth is called 'Ni Ch'i, adverse breath,' which is extremely harmful," states a passage from the Tao. "Be careful not to have the breath inhaled through the mouth."

As I lie in bed tossing and turning, fighting the urge to run to the bathroom again, I'm trying to focus on the positive, and am reminded of a skull from Marianna Evans's collection that offered a much-needed dose of hope.

. . .

It was morning, and Evans was seated in front of an oversize computer monitor in the administrative office of her orthodontics practice, about a half hour west of downtown Philadelphia. With white walls and white-tiled floors, the place looked futuristic. It was the opposite of the tan-stucco strip-mall blocks with ferns, goldfish tanks, and Robert Doisneau prints of all the dental offices I'd been to. Evans, I learned, ran a different kind of practice.

She brought up two images on a computer monitor, one of an ancient skull from the Morton Collection, and the other of a young girl, a new patient. I'll call her Gigi. Gigi was about seven in the photo. Her teeth jutted from the top of her gums, outward, inward, and in all directions. There were dark circles under her eyes; her lips were chapped and open as if she were sucking on an imaginary Popsicle. She suffered from chronic snoring, sinusitis, and asthma. She'd just started developing allergies to foods, dust, and pets.

Gigi grew up in a wealthy household. She followed the Food Guide Pyramid, got plenty of outdoor exercise, had her immunizations, took vitamins D and C, and had no illnesses growing up. And yet, here she was. "I see patients like this all day," said Evans. "They are all the same."

And here we are. Ninety percent of children have acquired some degree of deformity in their mouths and noses. Forty-five percent of adults snore occasionally when sleeping, and a quarter of the population snores constantly. Twenty-five percent of American adults over 30 choke on themselves because of sleep apnea; and an estimated 80 percent of moderate or severe cases are undiagnosed. Meanwhile, the majority of the population suffers from some form of breathing difficulty or resistance.

We've found ways to clean up our cities and to tame or kill off so many of the diseases that destroyed our ancestors. We've become more literate, taller, and stronger. On average, we live three times longer than people in the Industrial Age. There are now seven and a half billion humans on the planet—a thousand times more people than there were 10,000 years ago.

And yet we've lost touch with our most basic and important biological function.

Evans painted a depressing picture. And the irony wasn't lost on me as I sat in a sparkling clinic looking at one modern face after another and comparing them with the ideal form and perfect teeth of Samuel Morton's specimens, which he derided as "Australians and degraded Hottentots." At one point I scooted closer and saw my reflection in the monitor glass—that mangle of disjointed bones, the sloping jaw, stuffy nose, and mouth too small to fit all its teeth. *You fools,* I imagined that ancient skull saying. And for a moment, I swear, it looked like it was laughing.

But Evans hadn't invited me to see her research just to lament the present; her obsession with tracing the decline of human breathing is just a starting point. She'd studied it for years, entirely at her own expense, because she wants to help. She and her colleague, Kevin Boyd, are using the hundreds of measurements they've taken from ancient skulls to build a new model of airway health for modern humans. They are part of a burgeoning group of pulmonauts exploring novel therapies in breathing, lung expansion, orthodontics, and airway development. Their goal is to help return Gigi, me, and everyone else to our more perfect, ancient forms—the way we were before it all went haywire.

On the computer screen, Evans pulled up another photo. It was Gigi again, but in this shot there were no dark circles, none of the sallow skin or drooping lids. Her teeth were straight and her face was broad and glowing. She was nasal breathing again and no longer snored. Her allergies and other respiratory problems had all but disappeared. The photograph was taken two years after the first, and Gigi looked transformed.

The same thing happened with other patients—both adults and children—who'd regained the ability to breathe properly: their slack-jawed and narrowed faces morphed back into a more natural configuration. They saw their high blood pressure drop, depression abate, headaches disappear.

Harvold's monkeys recovered, too. After two years of forced mouthbreathing, he removed the silicone plugs. Slowly, surely, the animals relearned how to breathe through their noses. And slowly, surely, their faces and airways remodeled: jaws moved forward and facial structure and airways morphed back into their wide and natural state.

Six months after the experiment ended, the monkeys looked like monkeys again, because they were breathing normally again.

Back in my bedroom, staring out at the shadow play of branches in the window, I'm hoping that I too can reverse whatever damage I'd done in the last ten days, and the past four decades. I'm hoping I can relearn to breathe the way my ancestors breathed. I suppose I'll see soon enough.

Tomorrow morning, the plugs come out.

Part Two

o———o

THE LOST ART AND
SCIENCE OF BREATHING

Three

NOSE

○———○

"You look like shit," says Dr. Nayak.

It's early afternoon and I'm back at the Stanford Department of Otolaryngology Head and Neck Surgery Center. I'm splayed out on the examination chair while Nayak nudges an endoscope up my right nostril. The smooth desert dunes I journeyed through ten days ago look like they've been hit by a hurricane. I'll skip the details; let's just say my nasal cavity is a mess.

"Now your favorite part," says Nayak, chuckling. Before I sneeze or can consider running away, he grabs the wire brush and pushes it a few inches into my head. "It's pretty soupy in there," he says, sounding somewhat pleased. He repeats with the left nostril, places the gunk-covered RNA brushes into a test tube, then scoots me out of the way.

For the past week and a half I'd been waiting for this moment. I'd anticipated removing these plugs and tape and cotton to be a celebratory scene involving high-fives and nasal sighs of relief. I could breathe like a healthy human again!

In reality, it's minutes of discomfort followed by more obstruction. My nose is such a mess that Nayak has to grab a pair of pliers and insert several inches of cotton swabs into each nostril to keep whatever is up there from spilling onto the floor. Then it's back to the pulmonary function tests, an X-ray, the phlebotomist, and the rhinologist, repeating all the tests Olsson and I took before the obstruction phase. The results will be ready in a few weeks.

It's not until I get home that evening and rinse my sinuses several times that I can take a first full breath through my nose. I grab a coat and walk barefoot to the backyard. There are wispy plumes of cirrus clouds moving across the night sky, as big as spaceships. Above them, a few stubborn stars punch through the mist and cluster around a waxing moon.

I exhale stale air from my chest and take in a breath. I smell the sour, old-sock stink of mud. The black-label ChapStick of the damp doormat. A Lysol whiff of the lemon tree and the anise tinge of dying leaves.

Each of these scents, this material in the world, explodes in my head in a Technicolor burst. The scents are so sparkled and alarming that I can almost see them—a billion colored dots in a Seurat painting. As I take in another breath, I imagine all these molecules passing down my throat and into my lungs, pushing deeper into my bloodstream, where they provide fuel for thoughts and the sensations that made them.

Smell is life's oldest sense. Standing here alone, nostrils flaring, it occurs to me that breathing is so much more than just getting air into our bodies. It's the most intimate connection to our surroundings.

Everything you or I or any other breathing thing has ever put in its mouth, or in its nose, or soaked in through its skin, is hand-me-down space dust that's been around for 13.8 billion years. This wayward matter has been split apart by sunlight, spread throughout the universe, and come back together again. To breathe is to absorb ourselves in what surrounds us, to take in little bits of life, understand them, and give pieces of ourselves back out. Respiration is, at its core, reciprocation.

Respiration, I'm hoping, can also lead to restoration. Starting today, I will attempt to heal whatever damage has been done to my body over the past ten days of mouthbreathing and try to ensure ongoing health in the future. I'll put into practice several thousand years of teachings from several dozen pulmonauts, breaking down their methods and measuring the effects. Working with Olsson, I'll explore techniques to expand the lungs, develop the diaphragm, flood the body with oxygen, hack the autonomic nervous system, stimulate immune response, and reset chemoreceptors in the brain.

The first step is the recovery phase I've just done. To breathe through my nose, all day and all night.

The nose is crucial because it clears air, heats it, and moistens it for easier absorption. Most of us know this. But what so many people never consider is the nose's unexpected role in problems like erectile dysfunction. Or how it can trigger a cavalcade of hormones and chemicals that lower blood pressure and ease digestion. How it responds to the stages of a woman's menstrual cycle. How it regulates our heart rate, opens the vessels in our toes, and stores memories. How the density of your nasal hairs helps determine whether you'll suffer from asthma.

Few of us ever consider how the nostrils of every living person pulse to their own rhythm, opening and closing like a flower in response to our moods, mental states, and perhaps even the sun and the moon.

. . .

Thirteen hundred years ago, an ancient Tantric text, the *Shiva Swaro-daya*, described how one nostril will open to let breath in as the other will softly close throughout the day. Some days, the right nostril yawns awake to greet the sun; other days, the left awakens to the fullness of the moon. According to the text, these rhythms are the same throughout every month and they're shared by all humanity. It's a method our bodies use to stay balanced and grounded to the rhythms of the cosmos, and each other.

In 2004, an Indian surgeon named Dr. Ananda Balayogi Bhavanani attempted to scientifically test the *Shiva Swarodaya* patterns on an international group of subjects. Over the course of a month, he found that when the influence of the sun and moon on the Earth was at its strongest—during a full or new moon—the students consistently shared the *Shiva Swarodaya* pattern.

Bhavanani admitted the data were anecdotal and much more research would be needed to prove that all humans shared in this pattern. Still, scientists have known for more than a century that the nostrils *do* pulse to their own beat, that they do open and close like flowers throughout the day and night.

The phenomenon, called nasal cycles, was first described in 1895 by a German physician named Richard Kayser. He noticed that the tissue lining one nostril of his patients seemed to quickly congest and close while the other would mysteriously open. Then, after about 30 minutes to 4 hours, the nostrils switched, or "cycled." The shifting appeared to be influenced less by the moon's mysterious pull and more by sexual urges.

The interior of the nose, it turned out, is blanketed with erectile tissue, the same flesh that covers the penis, clitoris, and nipples. Noses get erections. Within seconds, they too can engorge with blood and become large and stiff. This happens because the nose is more intimately connected to the genitals than any other organ; when one gets aroused, the other responds. The mere thought of sex for some people causes such severe bouts of nasal erections that they'll have trouble breathing and will start to sneeze uncontrollably, an inconvenient condition called "honeymoon rhinitis." As sexual stimulation weakens and erectile tissue becomes flaccid, the nose will, too.

After Kayser's discovery, decades passed and nobody offered a good reason for why the human nose was lined with erectile tissue, or why the nostrils cycled. There were many theories: some believed this switching provoked the body to flip over from side to side while sleeping to prevent bedsores. (Breathing is easier through the nostril opposite the pillow.) Others thought the cycling helped protect the nose from respiratory infection and allergies, while still others argued that alternate airflow allows us to smell odors more efficiently.

What researchers eventually managed to confirm was that nasal erectile tissue mirrored states of health. It would become inflamed during sickness or other states of imbalance. If the nose became infected, the nasal cycle became more pronounced and switched back and forth quickly. The right and left nasal cavities also worked like an HVAC system, controlling temperature and blood pressure and feeding the brain chemicals to alter our moods, emotions, and sleep states.

The right nostril is a gas pedal. When you're inhaling primarily through this channel, circulation speeds up, your body gets hotter, and cortisol levels, blood pressure, and heart rate all increase. This happens because breathing through the right side of the nose activates the sympathetic nervous system, the "fight or flight" mechanism that puts the body in a more elevated state of alertness and readiness. Breathing through the right nostril will also feed more blood to the opposite hemisphere of the brain, specifically to the prefrontal cortex, which has been associated with logical decisions, language, and computing.

Inhaling through the left nostril has the opposite effect: it works as a kind of brake system to the right nostril's accelerator. The left nostril is more deeply connected to the parasympathetic nervous system, the rest-and-relax side that lowers blood pressure, cools the body, and reduces anxiety. Left-nostril breathing shifts blood flow to the opposite side of the prefrontal cortex, to the area that influences creative thought and plays a role in the formation of mental abstractions and the production of negative emotions.

In 2015, researchers at the University of California, San Diego, recorded the breathing patterns of a schizophrenic woman over the course of three consecutive years and found that she had a "significantly greater" left-nostril dominance. This breathing habit, they hypothesized, was likely overstimulating the right-side "creative part" of her brain, and as a result prodding her imagination to run amok. Over several sessions, the researchers taught her to breathe through her opposite, "logical" nostril, and she experienced far fewer hallucinations.

Our bodies operate most efficiently in a state of balance, pivoting between action and relaxation, daydreaming and reasoned thought. This balance is influenced by the nasal cycle, and may even be controlled by it. It's a balance that can also be gamed.

There's a yoga practice dedicated to manipulating the body's functions with forced breathing through the nostrils. It's called *nadi shodhana*—in Sanskrit, *nadi* means "channel" and *shodhana* means "purification"—or, more commonly, alternate nostril breathing.

I've been conducting an informal study of alternate nostril breathing for the past several minutes.

It's the second day of the nasal breathing "Recovery" phase, and I'm sitting in my living room, my elbows on the cluttered dining room table, softly sucking air through my right nostril, pausing for five seconds, and then blowing it out.

There are dozens of alternate nostril breathing techniques. I've started with the most basic. It involves placing an index finger over the left nostril and then inhaling and exhaling only through the right. I did this two dozen times after each meal today, to heat up my body and aid my digestion. Before meals, and any other time I wanted to relax, I'd switch sides, repeating the same exercise with my left nostril open. To gain focus and balance the body and mind, I followed a technique called *surya bheda pranayama*, which involves taking one breath into the right nostril, then exhaling through the left for several rounds.

These exercises felt great. Sitting here after a few rounds, I sense an immediate and potent clarity and relaxation, even a floatiness. As advertised, I've been entirely free of any gastroesophageal reflux. I haven't registered the slightest stomach ache. Alternate nostril breathing appeared to have delivered these benefits, but these techniques, I'd found, were usually fleeting, lasting only 30 minutes or so.

The real transformation in my body over the last 24 hours came from another practice: letting my nasal erectile tissues flex of their own accord, naturally adjusting the flow of air to suit the needs of my body and brain. It happened because of simply breathing through my nose.

As I'm quietly contemplating all this, Olsson comes barging in. "Good afternoon!" he yells. He's wearing his shorts and Abercrombie sweatshirt, and he plops down across from me while placing a blood-pressure cuff around his right arm. This is the same position he's assumed for the last eleven days straight, in pretty much the same clothes. Today, however, there's no bandage, nose clip, or silicone plugs up his nose. He's also breathing freely through his nostrils, taking in easy and silent inhales and exhales. His face is flushed, he's sitting upright, and he's so keyed up with energy that he can't stay still.

I figured that some of our new, bright outlook on life was psychosomatic until a few minutes later, when we checked our measurements. My systolic blood pressure had dropped from 142 ten days ago—a state of stage 2 hypertension—to 124, just a few points from a healthy range. My heart rate variability increased by more than 150 percent, and my carbon dioxide levels rose around 30 percent, placing me squarely within the medically normal zone. Olsson showed similar improvements.

To be clear, our little N=2 experiment proved little to nothing in the larger scope of scientific research. We were only two people, after all. At the same time, we can't dismiss the utter euphoria of returning to nose breathing, and the decades of real data from hundreds of real scientific studies that reached the same conclusion as our little experiment. That is: consciously controlling breathing can significantly influence our nervous system function, sleep quality, heartbeat, and blood flow.

And there's potential for much more. Because pulsing nasal cycles are only a small part of the nose's vital functions.

In a single breath, more molecules of air will pass through your nose than all the grains of sand on all the world's beaches—trillions and trillions of them. These little bits of air come from a few feet or several yards away. As they make their way toward you, they'll twist and spool like the stars in a van Gogh sky, and they'll keep twisting and spooling and scrolling as they pass into you, traveling at a clip of about five miles per hour.

What directs this rambling path are turbinates, six maze-like bones (three on each side) that begin at the opening of your nostrils and end just below your eyes. The turbinates are coiled in such a way that if you split them apart, they'd look like a seashell, which is how they got their other name, *nasal concha*, after the conch shell. Mollusks use their elaborately designed shells to filter impurities and keep invaders out. So do we.

The lower turbinates at the opening of the nostrils are covered in that pulsing erectile tissue, itself covered in mucous membrane, a nappy sheen of cells that moistens and warms breath to your body temperature while simultaneously filtering out particles and pollutants. All these invaders could cause infection and irritation if they got into the lungs; the mucus is the body's "first line of defense." It's constantly on the move, sweeping along at a rate of about half an inch every minute, more than 60 feet per day. Like a giant conveyor belt, it collects inhaled debris in the nose, then moves all the junk down the throat and into the stomach, where it's sterilized by stomach acid, delivered to the intestines, and sent out of your body.

This conveyor belt doesn't just move by itself. It's pushed along by millions of tiny, hair-like structures called cilia. Like a field of wheat in the wind, cilia sway with each inhale and exhale, but do so at a fast clip of up to 16 beats per second. Cilia closer to the nostrils gyrate at a different rhythm than those farther along, their movements creating a coordinated wave that keeps mucus moving deeper. The cilia grip is so strong that it can even push against the force of gravity. No matter what position the nose (and head) is in, whether it's upside down or right-side up, the cilia will keep pushing inward and down.

Working together, the different areas of the turbinates will heat, clean, slow, and pressurize air so that the lungs can extract more oxygen with each breath. This is why nasal breathing is far more healthy and efficient than breathing through the mouth. As Nayak explained when I first met him, the nose is the silent warrior: the gatekeeper of our bodies, pharmacist to our minds, and weather vane to our emotions.

. . .

The magic of the nose, and its healing powers, wasn't lost on the ancients.

Around 1500 BCE, the Ebers Papyrus, one of the oldest medical texts ever discovered, offered a description of how nostrils, not the mouth, were supposed to feed air to the heart and lungs. A thousand years later, Genesis 2:7 described how "the Lord God formed man of the dust of the ground, and breathed into his nostrils the breath of life; and man became a living soul." A Chinese Taoist text from the eighth century AD noted that the nose was the "heavenly door," and that breath must be taken in through it. "Never do otherwise," the text warned, "for breath would be in danger and illness would set in."

But it wasn't until the nineteenth century that the Western population ever considered the glories of nasal breathing. It happened thanks to an adventurous artist and researcher named George Catlin.

By 1830, Catlin had left what he called a "dry and tedious" job as a lawyer to become a portrait painter for Philadelphia's high society. He became well-known for his depictions of governors and aristocrats, but all the pomp and pretention of polite society did not impress him. Although his health was failing, Catlin yearned to be far away in nature, to capture rawer and more real depictions of humanity. He packed a gun, several canvases, a few paintbrushes, and headed west. Catlin would spend the next six years traveling thousands of miles throughout the Great Plains, covering more distance than Lewis and Clark to document the lives of 50 Native American tribes.

He went up the Missouri to live with the Lakota Sioux. He met with the Pawnee, Omaha, Cheyenne, and Blackfeet. Along the banks of the Upper Missouri, he happened upon the civilization of the Mandan, a mysterious tribe whose members stood six feet tall and lived in bubble-shaped houses. Many had luminous blue eyes and snow-white hair.

Catlin realized that nobody really knew about the Mandan, or other Plains tribes, because no one of European descent had bothered to spend time talking to them, researching them, living with them, and learning about their beliefs and traditions.

"I am traveling this country, as I have before said, not to advance or to prove *theories*, but to see all I am able to see and to tell it in the simplest and most intelligible manner I can to the world, for their own conclusions," Catlin wrote. He would paint some 600 portraits and take hundreds of pages of notes, forming what famed author Peter Matthiessen would call "the first, last, and only complete record ever made of the Plains Indians at the height of their splendid culture."

The tribes varied region by region, with different customs, traditions, and diets. Some, like the Mandan, ate only buffalo flesh and maize, while others lived primarily on venison and water, and still others harvested plants and flowers. The tribes looked different, too, with varying hair colors, facial features, and skin tones.

And yet Catlin marveled at the fact that all 50 tribes seemed to share the same superhuman physical characteristics. In some groups, such as the Crow and the Osage, Catlin wrote there were few men, "at their full growth, who are less than six feet in stature, and very many of them six and a half, and others seven feet." They all seemed to share a Herculean make of broad shoulders and barrel chests. The women were nearly as tall and just as striking.

Having never seen a dentist or doctor, the tribal people had teeth that were perfectly straight—"as regular as the keys of a piano," Catlin noted. Nobody seemed to get sick, and deformities and other chronic health problems appeared rare or nonexistent. The tribes attributed their vigorous health to a medicine, what Catlin called the "great secret of life." The secret was breathing.

The Native Americans explained to Catlin that breath inhaled through the mouth sapped the body of strength, deformed the face, and caused stress and disease. On the other hand, breath inhaled through the nose kept the body strong, made the face beautiful, and prevented disease. "The air which enters the lungs is as different from that which enters the nostrils as distilled water is different from the water in an ordinary cistern or a frog-pond," he wrote.

Healthy nasal breathing started at birth. Mothers in all these tribes followed the same practices, carefully closing the baby's lips with their fingers after each feeding. At night, they'd stand over sleeping infants and gently pinch mouths shut if they opened. Some Plains tribes strapped infants to a straight board and placed a pillow beneath their heads, creating a posture that made it much harder to breathe through the mouth. During winter, infants would be wrapped in light clothing and then held at arm's length on warmer days so they'd be less prone to get too hot and begin panting.

All these methods trained children to breathe through their noses, all day, every day. It was a habit they would carry with them the rest of their lives. Catlin described how adult tribal members would even resist smiling with an open mouth, fearing some noxious air might get in. This practice was as "old and unchangeable as their hills," he wrote, and it was shared universally throughout the tribes for millennia.

Twenty years after Catlin explored the West, he set off again, at age 56, to live with indigenous cultures in the Andes, Argentina, and Brazil. He wanted to know if "medicinal" breathing practices extended beyond the Plains. They did. Every tribe Catlin visited over the next several years—dozens of them—shared the same breathing habits. It was no coincidence, he reported, that they also shared the same vigorous health, perfect teeth, and forward-growing facial structure. He wrote about his experiences in *The Breath of Life*, published in 1862. The book was devoted solely to documenting the wonders of nasal breathing and the hazards of mouthbreathing.

Catlin was not only a chronicler of breathing methods; he was a practitioner. Nasal breathing saved his life.

As a boy, Catlin snored and was wracked with one respiratory problem after another. By the time he reached his 30s and first went out West, these problems had become so severe that he'd sometimes spit up blood.

His friends were convinced he had lung disease. Every night Catlin feared he would die.

"I became fully convinced of the danger of the habit [mouthbreathing], and resolved to overcome it," he wrote. Through "sternness of resolution and perseverance," Catlin forced his mouth closed while he slept and always breathed through his nose during waking hours. Soon, there were no more aches, pains, or bleeding. By his mid-30s, Catlin reported feeling healthier and stronger than at any other time in his life. "I at length completely conquered an insidious enemy that was nightly attacking me in my helpless position, and evidently fast hurrying me to the grave," he wrote.

George Catlin would live to be 76, about double the average life expectancy at the time. He credited his longevity to the "great secret of life": to *always* breathe through the nose.

.   .   .

It's the third night of the nasal breathing phase of the experiment, and I'm sitting up in bed reading, taking slow and easy breaths through my nose. I'm not breathing this way out of some "constant adult conviction," as Catlin wrote. I'm doing it because my lips are taped shut.

Catlin suggested tying a bandage around the jaw at night, but that sounded dangerous and difficult, so I opted for another technique, which I'd heard about months earlier from a dentist who runs a private practice in Silicon Valley.

Dr. Mark Burhenne had been studying the links between mouthbreathing and sleep for decades, and had written a book on the subject. He told me that mouthbreathing contributed to periodontal disease and bad breath, and was the number one cause of cavities, even more damaging than sugar consumption, bad diet, or poor hygiene. (This belief had been echoed by other dentists for a hundred years, and was endorsed by Catlin too.) Burhenne also found that mouthbreathing was both a cause of and a contributor to snoring and sleep apnea. He recommended his patients tape their mouths shut at night.

"The health benefits of nose breathing are undeniable," he told me. One of the many benefits is that the sinuses release a huge boost of nitric oxide, a molecule that plays an essential role in increasing circulation and delivering oxygen into cells. Immune function, weight, circulation, mood, and sexual function can all be heavily influenced by the amount of nitric oxide in the body. (The popular erectile dysfunction drug sildenafil, known by the commercial name Viagra, works by releasing nitric oxide into the bloodstream, which opens the capillaries in the genitals and elsewhere.)

Nasal breathing alone can boost nitric oxide sixfold, which is one of the reasons we can absorb about 18 percent more oxygen than by just breathing through the mouth. Mouth taping, Burhenne said, helped a five-year-old patient of his overcome ADHD, a condition directly attributed to breathing difficulties during sleep. It helped Burhenne and his wife cure their own snoring and breathing problems. Hundreds of other patients reported similar benefits.

The whole thing seemed a little sketchy until Ann Kearney, a doctor of speech-language pathology at the Stanford Voice and Swallowing Center, told me the same. Kearney helped rehabilitate patients who had swallowing and breathing disorders. She swore by mouth taping.

Kearney herself had spent years as a mouthbreather due to chronic congestion. She visited an ear, nose, and throat specialist and discovered that her nasal cavities were blocked with tissue. The specialist advised that the only way to open her nose was through surgery or medications. She tried mouth taping instead.

"The first night, I lasted five minutes before I ripped it off," she told me. On the second night, she was able to tolerate the tape for ten minutes. A couple of days later, she slept through the night. Within six weeks, her nose opened up.

"It's a classic example of use it or lose it," Kearney said. To prove her claim, she examined the noses of 50 patients who had undergone laryngectomies, a procedure in which a breathing hole is cut into the throat. Within two months to two years, every patient was suffering from complete nasal obstruction.

Like other parts of the body, the nasal cavity responds to whatever inputs it receives. When the nose is denied regular use, it will atrophy. This is what happened to Kearney and many of her patients, and to so much of the general population. Snoring and sleep apnea often follow.

Keeping the nose constantly in use, however, trains the tissues inside the nasal cavity and throat to flex and stay open. Kearney, Burhenne, and so many of their patients healed themselves this way: by breathing from their noses, all day and all night.

How to apply mouth tape, or "sleep tape" as it's also called, is a matter of personal preference, and everyone I talked to had their own technique. Burhenne liked to place a small piece horizontally over the lips; Kearney preferred a fat strip over the entire mouth. The internet was filled with suggestions. One guy used eight pieces of inch-wide tape to create a sort of tape goatee. Another used duct tape. A woman suggested taping the entire lower half of the face.

To me, these methods are ridiculous and excessive. Looking for an easier way, over the last few days I conducted my own experiments with blue painter's tape, which smelled weird, and Scotch tape, which crinkled. Band-Aids were too sticky.

Eventually I realized that all I or anyone really needed was a postage-stamp-size piece of tape at the center of the lips—a Charlie Chaplin mustache moved down an inch. That's it. This approach felt less claustrophobic and allowed a little space on the sides of the mouth if I needed to cough or talk. After much trial and error, I settled on 3M Nexcare Durapore "durable cloth" tape, an all-purpose surgical tape with a gentle adhesive. It was comfortable, had no chemical scent, and didn't leave residue.

In the three nights since I started using this tape, I went from snoring four hours to only ten minutes. I'd been warned by Burhenne that sleep tape won't do anything to help treat sleep apnea. My experience suggested otherwise. As my snoring disappeared, so did apnea.

I'd suffered up to two dozen apnea events in the mouthbreathing phase, but last night had zero. I suffered no creepy insomniac hallucinations, no late-night ruminations on *Homo habilis* or Edward Gorey. I never woke up needing to pee. I didn't have to, because my pituitary gland was likely releasing vasopressin. I was finally sleeping soundly.

Meanwhile, Olsson went from snoring half the night to not snoring for even a minute. His apnea events dropped from 53 to zero. The bright-eyed, cotton-haired Swede I'd felt so guilty about abusing had been reborn. Earlier today he was smiling, so convinced of sleep tape's healing power that he kept a piece of it stuck to his lips for the rest of the morning.

Sleep, and life, had become something that Olsson and I embraced again. Now, sitting in bed, with a little stamp of white tape stuck to my lips, I flipped to the last page in Catlin's *Breath of Life*, the final paragraph he'd ever publish in his long life of research.

"And if I were to endeavor to bequeath to posterity the most important Motto which human language can convey, it should be in three words—SHUT-YOUR-MOUTH*. . . . Where I would paint and engrave it, in every *Nursery*, and on every *Bed-post* in the Universe, its meaning could not be mistaken.

"And if obeyed," he continued, "its importance would soon be realized."

---

*"Shutting your mouth" means both inhaling and exhaling through the nose as much as possible. It turns out that exhaling (nasal and otherwise) is a therapy unto itself, as you'll soon discover.

Four

EXHALE

Every morning at 9:00, after Olsson and I have finished our testing and split off for solo time, I roll out a mat on my living room floor and work on becoming a little more immortal.

The path to everlasting life involves a lot of stretching: back bends, neck bends, and twirling, each one a holy and ancient practice that had been passed down in secrecy from one Buddhist monk to another for 2,500 years. Olsson and I need this stretching; even if we breathe through the nose twenty-four hours a day, it won't help much unless we've got the lung capacity to hold in that air. Just a few minutes of daily bending and breathing can expand lung capacity. With that extra capacity we can expand our lives.

The stretches, called the Five Tibetan Rites, came to the Western world, and to me, by way of writer Peter Kelder, who was known as a lover of "books and libraries, words and poetry."

In the 1930s, Kelder was sitting on a park bench in southern California when an elderly stranger struck up a conversation. The man, whom he called Colonel Bradford, had spent decades in India with the British Army. The Colonel was old—all sloping shoulders, gray hair, and wobbly legs—but he believed there was a cure for aging and that it was locked up in a monastery in the Himalayas. The usual mystical stuff occurred up there: the sick became healthy, poor became rich, old became young. Kelder and the Colonel stayed in touch and shared many conversations. Then, one day, the old man hobbled away, desperate to find this Shangri-La before he drew his last breath.

Four years passed until Kelder received a call from his building's doorman. The Colonel was waiting downstairs. He looked 20 years younger. He was standing straight, his face vibrant and alive, and his once-balding head was covered in thick, dark hair. He'd found the monastery, studied the ancient manuscripts, and learned restorative practices from the monks. He'd reversed aging through nothing more than stretching and breathing.

Kelder described these techniques in a slim booklet titled *The Eye of Revelation,* published in 1939. Few people bothered to read it; fewer believed it. Kelder's yarn was likely fabricated, or at minimum grossly exaggerated. However, the lung-expanding stretches he described are rooted in actual exercises that date back to 500 BCE. Tibetans had used these methods for millennia to improve physical fitness, mental health, cardiovascular function, and, of course, extend life.

More recently, science has begun measuring what the ancient Tibetans understood intuitively. In the 1980s, researchers with the Framingham Study, a 70-year longitudinal research program focused on heart disease, attempted to find out if lung size really did correlate to longevity. They gathered two decades of data from 5,200 subjects, crunched the numbers, and discovered that the greatest indicator of life span wasn't genetics, diet, or the amount of daily exercise, as many had suspected. It was lung capacity.

The smaller and less efficient lungs became, the quicker subjects got sick and died. The cause of deterioration didn't matter. Smaller meant shorter. But larger lungs equaled longer lives.

Our ability to breathe full breaths, according to the researchers, appears to be "literally a measure of living capacity." In 2000, University of Buffalo researchers ran a similar study, comparing lung capacity in a group of more than a thousand subjects over three decades. The results were the same.

What neither of these landmark studies addressed, however, was how a person with deteriorated lungs might heal and strengthen them. There were surgeries to remove diseased tissue and drugs to stem infections, but no advice on how to keep lungs large and healthy throughout life. All the way up to the 1980s, the common belief in Western medicine was that the lungs, like every other internal organ, were immutable. That is, whatever lungs we were born with, we were stuck with. As these organs degraded with age, the only thing we could do was sigh and bear it.

Aging was supposed to go like this: Starting around 30, we should expect to lose a little more memory, mobility, and muscle with every passing year. We would also lose the ability to breathe properly. Bones in the chest would become thinner and change shape, causing rib cages to collapse inward. Muscle fibers surrounding the lungs would weaken and prevent air from entering and exiting. All these things reduce lung capacity.

The lungs themselves will lose about 12 percent of capacity from the age of 30 to 50, and will continue declining even faster as we get older, with women faring worse than men. If we make it to 80, we'll be able to take in 30 percent less air than we did in our 20s. We're forced to breathe faster and harder. This breathing habit leads to chronic problems like high blood pressure, immune disorders, and anxiety.

But what the Tibetans have long known and what Western science is now discovering is that aging doesn't have to be a one-way path of decline. The internal organs are malleable, and we can change them at nearly any time.

Freedivers know this better than anyone. I'd learned it from them years ago, when I met several people who had increased their lung capacity by an astounding 30 to 40 percent. Herbert Nitsch, a multiple world record holder, reportedly has a lung capacity of 14 liters—more than *double* that of the average male. Neither Nitsch nor any of the other freedivers started out like this; they made their lungs larger by force of will. They taught themselves how to breathe in ways that dramatically changed the internal organs of their bodies.

Fortunately, diving down hundreds of feet is not required. Any regular practice that stretches the lungs and keeps them flexible can retain or increase lung capacity. Moderate exercise like walking or cycling has been shown to boost lung size by up to 15 percent.

These discoveries would have been welcome news to Katharina Schroth, a teenager who lived in Dresden, Germany, in the early 1900s. Schroth had been diagnosed with scoliosis, a sideways curvature of the spine. The condition had no cure, and most children who suffered from extreme cases like Schroth's could expect to spend a life in bed or rolling around in a wheelchair.

Schroth had other thoughts about the human body's potential. She'd watched how balloons collapsed or expanded, pushing or pulling in whatever was around them. The lungs, she felt, were no different. If she could expand her lungs, maybe she could also expand her skeletal structure. Maybe she could straighten her spine and improve the quality and quantity of her life.

At age 16, Schroth began training herself in something called "orthopedic breathing." She would stand in front of a mirror, twist her body, and inhale into one lung while limiting air intake to the other. Then she'd hobble over to a table, sling her body on its side, and arch her chest back and forth to loosen her rib cage while breathing into the empty space. Schroth spent five years doing this. At the end, she'd effectively cured herself of "incurable" scoliosis; she'd breathed her spine straight again.

Schroth began teaching the power of breathing to other scoliosis patients, and by the 1940s she was running a bustling institute in rural western Germany. It had no hospital rooms or other standard medical equipment; just a few run-down buildings, a yard, a picket fence, and patio tables. One hundred fifty scoliosis patients would gather there at a time. They suffered from the most severe form of the disease, with spines curved more than 80 degrees. Many were so hunched over, their backs so twisted and turned, that they couldn't walk or even look upward. Their disfigured ribs and chests made it hard for them to take a breath, and they likely suffered from respiratory problems, fatigue, and heart conditions because of it. Hospitals had given up on trying to heal these patients. They came to live with Schroth for six weeks.

The German medical community derided Schroth, claiming that she was neither a professional trainer nor a physician and was not qualified to treat patients. She ignored them all; she kept doing things her way, having the women strip down bare-chested in a dirt lot beneath a copse of beech trees, stretching and breathing themselves back to health. Within a few weeks, hunched backs straightened and many students gained inches in height. Women who had been bedridden and hopeless finally began to walk again. They could take full breaths again.

Schroth spent the next 60 years bringing her techniques to

hospitals throughout Germany and beyond. Toward the end of her life, the medical community had changed its tune and the German government awarded Schroth the Federal Cross of Merit for her contributions to medicine.

"What the bodily form depends on is breath (chi) and what breath relies upon is form," states a Chinese adage from 700 AD. "When the breath is perfect, the form is perfect (too)."

Schroth continued to expand her lungs and improve her own breathing and form throughout her life. This former scoliosis patient, who as a teenager had been left to wither in bed, would die in 1985, just three days shy of her 91st birthday.

. . .

Midway through my research for this book, I took a trip to New York City to meet a more contemporary breathing expert who offered a different approach to expanding the lungs and longevity. Her apartment workspace was located a few blocks from the United Nations in a brown brick building with an awning covered in pink-eyed pigeons. I walked past a drowsy doorman, rode up an elevator, and a minute later I was knocking at room 418.

Lynn Martin welcomed me in. She was beanpole thin and dressed in a black jumpsuit with an oversize brass-buckled belt. "I told you it was small!" she said of the studio apartment. Surrounding us were manila folders, human anatomy books, and a few plastic models of human lungs. On a wall beside a bookshelf were black-and-white photos of Martin in the early 1970s. In one, she was wearing a black leotard in mid-glide on the wood floor of a dance studio, her blond hair pulled back in a lazy ponytail, her face bearing an uncanny resemblance to *Rosemary's Baby*–era Mia Farrow.

After a few pleasantries, Martin sat me down and started telling me what I came to hear. "He was very verbal, but when you asked what exactly he was doing, he could never explain it," she said. "Nobody since has ever been able to do what he did."

The subject of intrigue was Carl Stough, a choir conductor and medical anomaly who got his start in the 1940s. Of all the pulmonauts I'd come across over the past several years, Stough was the most elusive. He published one book in 1970, which quickly flopped and went out of print. Twenty years later, a CBS producer put together a one-hour program about his groundbreaking work, but it never aired. Stough himself didn't advertise his techniques. He never went on speaking tours. Even so, professional opera singers, Grammy-winning saxophonists, paraplegics, and dying emphysemics—thousands of them—managed to find him. Stough broke all the rules; he expanded lungs and extended life spans. And yet, most people today have never heard of him.

Martin had worked with Stough for more than two decades. She was a living link to this mysterious man and his research in the lost art of breathing. What Stough had discovered, and what Martin had learned, was that the most important aspect of breathing wasn't just to take in air through the nose. Inhaling was the easy part. The key to breathing, lung expansion, and the long life that came with it was on the other end of respiration. It was in the transformative power of a full exhalation.

Photographs of Stough from the 1940s show an upright man who bore a passing resemblance to Thurston Howell III, the millionaire from *Gilligan's Island*. Stough liked to sing and teach singing. He noticed how his fellow singers would belt out a few measures, stop to take a breath, and then belt out a few more. Each seemed to be gasping for air, holding it high in the chest, and releasing it too soon. Singing, talking, yawning, sighing—any vocalization we make occurs during the exhalation. Stough's students had thin and weak voices because, he believed, they had thin and weak exhalations.

While directing choirs at Westminster Choir College in New Jersey, Stough began training his singers to exhale properly, to build up their respiratory muscles and enlarge their lungs. Within a few sessions, the students were singing clearer, more robustly, and with added nuance. He moved to North Carolina to conduct church choirs that went on to win national competitions, and his choir appeared on a weekly program broadcast nationally by Liberty Radio Network. Stough became so renowned that he moved to New York to retrain singers at the Metropolitan Opera.

In 1958, the administration of the East Orange Veterans Affairs Hospital in New Jersey called. "You must know something about breathing that we don't," said Dr. Maurice J. Small, the chief of tuberculosis management. Small was wondering if Stough might be interested in training a new group of students. None of them could sing, and a few couldn't walk or talk. They were emphysema patients, and they were in desperate need of help.

When Stough arrived at the East Orange hospital weeks later, he was horrified. Dozens of patients were laid out on gurneys, each one jaundiced and pale, their mouths craned open like fish, oxygen tubes pumping to no avail. The hospital staff didn't know what to do, so they just wheeled the men across the waxed terrazzo floors and into a room hung with faded-yellow tissue dispensers and American flag clocks, one patient beside the other, waiting to die. It had gone this way for 50 years.

"I foolishly had assumed that everyone had at least a rudimentary knowledge of physiology," Stough wrote in his autobiography, *Dr. Breath*. "Even more foolishly I had assumed that a universal awareness of the importance of breathing existed. Nothing could have been farther from the truth."

Emphysema is a gradual deterioration of lung tissue marked by chronic bronchitis and coughing. The lungs become so damaged that people with the disease can no longer absorb oxygen effectively. They're forced to take several short breaths very fast, often breathing in far more air than they need, but still feel out of breath. Emphysema had no known cure.

The nurses, meaning well, had placed cushions under patients' backs so that their chests were arched up. The idea was to create elevation to ease inhalations. Stough instantly saw that this was making the condition worse.

Emphysema, he realized, was a disease of exhalation. The patients were suffering not because they couldn't get fresh air into their lungs, but because they couldn't get enough stale air out.

Normally, the blood coursing through our arteries and veins at any one time does a full circuit once a minute, an average of 2,000 gallons of blood a day. This regular and consistent blood flow is essential to delivering fresh oxygenated blood to cells and removing waste.

What influences much of the speed and strength of this circulation is the thoracic pump, the name for the pressure that builds inside the chest when we breathe. As we inhale, negative pressure draws blood into the heart; as we exhale, blood shoots back out into the body and lungs, where it recirculates. It's similar to the way the ocean floods into shore, then ebbs out.

And what powers the thoracic pump is the diaphragm, the muscle that sits beneath the lungs in the shape of an umbrella. The diaphragm lifts during exhalations, which shrinks the lungs, then it drops back down to expand them during inhalations. This up-and-down movement helps to stabilize the spine and occurs within us some 50,000 times a day.

A typical adult engages as little as 10 percent of the range of the diaphragm when breathing, which overburdens the heart, elevates blood pressure, and causes a rash of circulatory problems. Extending those breaths to 50 to 70 percent of the diaphragm's capacity will ease cardio-vascular stress and allow the body to work more efficiently. For this reason, the diaphragm is sometimes referred to as "the second heart," because it not only beats to its own rhythm but also affects the rate and strength of the heartbeat.

Stough discovered that the diaphragms in all of the East Orange emphysema patients had broken down. X-rays showed that they were extending their diaphragms by only a fraction of what was healthy, taking only a sip of air with each breath. The patients had been sick so long that many of the muscles and joints around their chests had atrophied and stiffened; they had no muscle memory of breathing deep. Over the next two months, Stough reminded them how.

"My activities looked silly when observed from a distance, and they seemed silly at the outset to the person with whom I worked," wrote Stough.

He'd begin the treatments by putting patients flat on their backs, running his hands across their torsos, and gently tapping on rigid muscles and distended chests. He'd have them hold their breath and count from one to five as many times in a row as they could. Next, he massaged their necks and throats and lightly coaxed their ribs as he told them to inhale and exhale *very slowly*, trying to wake the diaphragm from its long slumber. Each of these exercises allowed the patients to let out a little more air so that a little more air could get in.

After several sessions, some patients learned to speak a full sentence in a single breath for the first time in years. Others began walking.

"One elderly man who had not been able to walk across the room not only could walk but could walk up the hospital stairs, a remarkable feat for an advanced emphysema patient," Stough wrote. Another man, who hadn't been able to breathe for more than 15 minutes without supplemental oxygen, was lasting eight hours. A 55-year-old man who had suffered advanced emphysema for eight years was able to leave the hospital and captain a boat down to Florida.

Before-and-after X-rays showed that Stough's patients were vastly expanding their lung capacity in only a few weeks. Even more stunning, they were training an involuntary muscle—the diaphragm—to lift higher and drop lower. Administrators told Stough that this was medically impossible; internal organs and muscles cannot be developed, they said. At one point, several doctors petitioned to ban Stough from treating patients and kick him out of the hospital system. Stough was a choral teacher, not a doctor, after all. But the X-rays didn't lie. To confirm his results, Stough began recording the first footage of a moving diaphragm, using a new X-ray film technology called cinefluorography. Everyone was floored.

"I told Carl in no uncertain words that he was mildly demented to say that he could effect a rise in the diaphragm and a descent in the ribs, but then in one patient we got rather spectacular results showing that he did do this," said Dr. Robert Nims, the chief of pulmonary medicine at the West Haven VA Hospital in Connecticut. "We have shown that he's able to decrease the volume of the lungs [via deep exhalations] more than any pulmonary man would say it was possible."

Stough hadn't found a way to reverse emphysema. Lung damage from the disease is permanent. What he'd done is find a way to access the rest of the lungs, the areas that were still functioning, and engage them on a larger level. The "cure" Stough professed was de facto, but it worked.

Over the next decade, Stough would take his treatment to a half-dozen of the largest VA hospitals on the East Coast, sometimes working on patients seven days a week. He'd go on to treat not only emphysema, but asthma, bronchitis, pneumonia, and more.

The benefits of breathing, of harnessing the art of exhalation, Stough found, extended not just to the chronically sick or to singers, but to everyone.

. . .

Back in Lynn Martin's apartment, I was reawakening my own slumbering diaphragm on the living room futon. "This is not a massage," Martin said, making her point clear as she pressed a hand against my ribs. I drew soft and long breaths deep into my gut while Martin helped loosen my rib cage, trying to coax at least 50 percent of my maximum diaphragm movement with each inhale and exhale.

Breathing this way wasn't necessary, Martin told me. Our bodies can survive on short and clipped breaths for decades, and many of us do. That doesn't mean it's good for us. Over time, shallow breathing will limit the range of our diaphragms and lung capacity and can lead to the high-shouldered, chest-out, neck-extended posture common in those with emphysema, asthma, and other respiratory problems. Fixing this breathing and this posture, she told me, was relatively easy.

After several rounds of deep breaths to open my rib cage, Martin asked me to start counting from one to ten over and over with every exhale. "*1, 2, 3, 4, 5, 6, 7, 8, 9, 10; 1, 2, 3, 4, 5, 6, 7, 8, 9, 10*—then keep repeating it," she said. At the end of the exhale, when I was so out of breath I couldn't vocalize anymore, I was to keep counting, but to do so silently, letting my voice trail down into a "sub-whisper."

I ran through a few rounds, counting quickly and loudly, then silently mouthing the numbers. At the end of each breath, it felt like my chest had been plastic-wrapped and my abs had just gone through a brutal workout. "Keep going!" Martin said.

The strain of the counting exercise is equivalent to the strain on the lungs during physical exertion. This was what made the exercise so effective for Stough's bedridden patients. The point was to get the diaphragm accustomed to this wider range so that deep and easy breathing became unconscious. "Keep moving your lips!" Martin egged me on. "Get out the last little molecule of air!"

After a few more minutes of counting, silent and otherwise, I stopped and took a break and felt my diaphragm chugging away like a piston in slow motion, radiating fresh blood from the center of my body. This is the feeling of what Stough called "Breathing Coordination," when the respiratory and circulatory systems enter a state of equilibrium, when the amount of air that enters us equals the amount that leaves, and our bodies are able perform all their essential functions with the least exertion.

In 1968, Stough left the VA system and his thriving private practice in New York to train yet another group of students. These people could talk, and they could walk, and they could run very fast. They were the runners on the Yale track and field team, among the best in the nation at the time. When Stough arrived at the track fieldhouse, the athletes were so excited that they hung a poster on the bulletin board outside: *Dr. Breath Is Here Today!*

Stough had expected these elite athletes to have exemplary breathing habits. Instead, he found that they suffered from the same "respiratory weakness" as everyone else: they got the same colds and flus and lung infections. Most of them breathed way too often, high in their chests. Sprinters were the worst off. The short and violent breaths they took during runs put too much pressure on delicate tissues and bronchial tubes. As a result, they suffered from asthma and other respiratory ailments. At the finish line, they coughed and sometimes vomited and collapsed, wheezing in pain.

"I had observed that in recovering from performance, athletes tended to adopt the same breathing characteristics as those the emphysema patients exhibited," Stough wrote. These runners had been trained to push through the pain, and they did. They won competitions, but they were harming their bodies.

Stough laid out a table at the Yale indoor track, sat the runners on it, and began running his hands up and around their chests in front of a crowd of onlookers. He warned them to never hold their breath when positioned at the starting line at the beginning of a race, but to breathe deeply and calmly and always exhale at the sound of the starter pistol. This way, the first breath they'd take in would be rich and full and provide them with energy to run faster and longer.

After only a few sessions, all the runners reported feeling better and breathing better. "I never felt so relaxed in my life," one sprinter said. They took half the time to recover between races and were soon breaking personal bests and edging toward world records.

On the heels of the Yale success, Stough moved to South Lake Tahoe to train runners preparing for the 1968 Summer Olympics in Mexico City. Same therapy, same success. A decathlete went out to the track and broke his previous record. Another broke his lifetime record. A runner named Rick Sloan broke his two life records for three events.

"Through my work with Dr. Stough, I knew I had to exhale," said Lee Evans, an Olympic sprinter. "You know, I exhale, which kept my energy up. I didn't get tired. . . . But after the game, I found that this was for my life."

You might recognize Evans. He's the man in the famous photograph standing on the center podium at the Olympics awards ceremony, wearing a Black Panther beret and jutting a fist in the air. He won gold in the 400 meters and another in the 400-meter relay. The rest of the 1968 U.S. men's team under Stough's training went on to win a total of 12 Olympic medals, most gold, and set five world records. It was one of the greatest performances in an Olympics. The Americans were the only runners to not use oxygen before or after a race, which was unheard of at the time.

They didn't need to. Stough had taught them the art of breathing coordination, and the power of harnessing a full exhalation.

"He was doing so many things at once," Lynn Martin said as we moved from the futon back to the dining room table at the center of her studio apartment. "The sensitivity of his hands, perfect pitch of his ears, the natural knack for instruction—all of it." For the past few minutes, Martin had been telling me about her time working with Stough, how she'd gone to see him in 1975 on the recommendation of another dancer and had come out feeling transformed. She returned weeks later and took a job at the clinic. Even though Martin would spend more than two decades working with Stough as one of his closest associates, he never told her his secrets. "He thought it was too difficult to put into words," she said.

I could relate. I'd seen a video recording of Stough at the 1992 Aspen Music Festival—the only existing footage that demonstrates what he did and how he did it. It opened with a frame that read: *An Introduction to Respiratory Science: The Preventative Medicine of the Twenty-First Century*. Stough was at the center of a conference room, a massage table in front of him. An open window looked out over a thicket of pine trees glowing white in the summer sun. Stough was deeply tanned and dressed in a black blazer with brass buttons and a pocket kerchief, as if he'd just flown in on a Concorde from Monte Carlo.

He started off by inviting a tenor named Timothy Jones to lie on the table and proceeded to jiggle Jones's jaw, dig his hands into his waist, and rock him back and forth. "You see, I have to keep tapping right on the chest," said Stough, his yellow polka-dot tie dangling in Jones's hair. This went on for several minutes, until Stough leaned three inches from Jones's face and began to count from one to ten with him in a gibberishy harmony. "Everything's loosening very fast!" Stough announced. He wiggled Jones's hips and neck so violently that the singer almost fell off the table.

It was a bizarre spectacle, and the grabbing and pushing and deep stroking looked at times like borderline molestation. After my own experience in Martin's studio for an hour, babbling numbers and having my chest poked and ribs squeezed, it became more clear to me why

Stough's work never caught on. It didn't matter that saxophonist David Sanborn and asthmatic opera singers, Olympic runners, and hundreds of emphysema survivors praised his treatments as a lifesaver. Stough wasn't a doctor; he was a self-made pulmonaut, a choir conductor. He was just too far out there. His therapy was just too weird.

"Although the process of breathing involves both anatomy and physiology, neither branch of science has claimed it for thorough exploration," wrote Stough. "It was a little-known territory waiting to be mapped and charted."

Stough made his map over a half century of constant work. But when he died, that map was lost. As soon as he left the VA wards, so did his therapy.

. . .

At the end of my two-hour Breathing Coordination session, I left Martin's apartment and hopped on the train back to Newark Liberty International Airport. As we rumbled across the marshlands and over the Passaic River, I searched through the current treatments for the nearly 4 million Americans now suffering from emphysema. There were bronchodilators, steroids, and antibiotics. There was supplemental oxygen and surgery and something called pulmonary rehabilitation, which included assistance to quit smoking, exercise planning, nutrition counseling, and some pursed-lip breathing techniques.

But there was no mention of Stough, or the "second heart" of the diaphragm, or the importance of a full exhalation. No mention of how expanding the lungs and breathing properly had effectively reversed the disease or lengthened lives. Emphysema was still listed as an incurable condition.

Five

SLOW

o———o

"Could you hand me the oximeter?" Olsson asks from across the dining room table. It's afternoon on the fifth day of the Recovery phase, and for the past 30 minutes we've been testing our pH levels, blood gases, heart rates, and other vital signs. This is the 45th time we've been through this drill in the past two weeks.

Although both Olsson and I feel utterly transformed while nasal breathing, the monotony of the days is becoming maddening. We're eating the same food at the same time as we did ten days earlier, sweating through the same stationary bike workouts in the same gym, and having many of the same conversations. This afternoon we're discussing Olsson's favorite subject, his obsession for the past decade. We are, once again, talking about carbon dioxide.

It is hard to admit now, but when I first interviewed Olsson more than a year ago, he was not a source I entirely trusted. On our Skype calls, he liked to hammer the importance of slow breathing, and he'd sent me a half-dozen PowerPoint presentations and reams of scientific studies on how paced breathing relaxed the body and calmed the mind. This part made perfect sense. But when he started in on the restorative wonders of a toxic gas, I began to wonder. "I really think carbon dioxide is more important than oxygen," he told me.

Olsson claimed that we have 100 times more carbon dioxide in our bodies than oxygen (which is true), and that most of us need even more of it (also true). He said it wasn't just oxygen but huge quantities of carbon dioxide that fostered the burst of life during the Cambrian Explosion 500 million years ago. He said that, today, humans can increase this toxic gas in our bodies and sharpen our minds, burn fat, and, in some cases, heal disease.

After a while I began to worry that Olsson was nuts, or at least prone to gross exaggeration, and that our hours of conversations had been a waste of time.

Carbon dioxide, after all, is a metabolic waste product. It's the stuff that plumes out of coal plants and rotten fruit. The instructor at a boxing class I attended used to beseech students to "breathe deep and get all that carbon dioxide out of your system." This seemed like good advice. Every other day, a new headline detailed how Earth was warming because there was too much carbon dioxide in the atmosphere. Animals were dying. Carbon dioxide kills.

Olsson kept arguing the opposite. He insisted that carbon dioxide could be beneficial, and he warned me that too much oxygen in my body wouldn't help me but hurt me. "Breathing heavy, breathing quickly and as deeply as you can—I realized this is the worst advice anyone could give you," Olsson told me. Big, heavy breaths were bad for us because they depleted our bodies of, yes, carbon dioxide.

Several months of this back-and-forth got me intrigued enough, or confused enough, or both, that I decided to fly to Sweden, spend a few days with Olsson, and see his operation in an attempt to learn more about one of the most misunderstood gases in the universe.

.   .   .

I arrived in Stockholm in mid-November and took a train to an industrial co-working space on the outskirts of town. Through the windows of a cavernous lobby, the sunlight had a kind of slant to it. Ominous clouds gathered, and the air was thick with the heavy feeling that precedes a long winter.

Olsson showed up exactly on time, took a seat across from me, and placed a glass of water on the table. He wore faded jeans, white tennis shoes, and a pressed white shirt. He had the kind of calm you see in monks, Amish, and others who spend a lot of time in their inner worlds. When he spoke, it was always softly, and with that annoying habit that all Scandinavians seem to have inherited: flawless English, with no *umm*s or *huh*s or pauses. He'd even gotten his *whom*s down and inserted the oft-forgotten "not" when he told me how he could care less.

"I was going to end up exactly like my father," Olsson said, running a finger along the condensation of the water glass. He told me how his father had been chronically stressed, how he breathed too much, and how he'd gotten severe high blood pressure and lung disease and died at 68 with a breathing tube in his mouth. "I knew that so many other people were going to stay sick and die of the same thing," Olsson explained. He wanted to educate himself so he'd be prepared if something else happened to him or his family.

After the long days he spent running a software distribution company, he'd come home and read medical books. He talked to doctors, surgeons, instructors, and research scientists. Eventually he sold his business, got rid of his nice cars and big house, got a divorce, and moved into a condominium. Then he scaled down to a smaller apartment and spent six years forgoing any salary, working almost entirely alone, trying to understand the mysteries of health, medicine, and most specifically breathing and the role of carbon dioxide in the body. "There were yogi books about prana and then there were medical books focusing on pathologies—blood gases and disease and CPAP," he said.

In short, Olsson found what I'd found, but years earlier: that there was a gap in our knowledge about the science of breathing and its role in our bodies. He discovered that we'd done a good job of examining what causes breathing problems but done little to explore how they first develop and how we might prevent them.

Olsson was in good company. Doctors had been complaining about this for decades. "The field of respiratory physiology is expanding in all directions, yet so preoccupied have most physiologists been with lung volumes, ventilation, circulation, gas exchange, the mechanics of breathing, the metabolic cost of breathing and the control of breathing that few have paid much attention to the muscles that actually do the breathing," one physician wrote in 1958. Another wrote: "Until the seventeenth century most of the great physicians and anatomists were interested in the respiratory muscles and the mechanics of breathing. Since then these muscles have been increasingly neglected, lying as they do in a no-man's land between anatomy and physiology."

What many of these doctors found, and what Olsson would discover much later, was that the best way to prevent many chronic health problems, improve athletic performance, and extend longevity was to focus on how we breathed, specifically to balance oxygen and carbon dioxide levels in the body. To do this, we'd need to learn how to inhale and exhale slowly.

. . .

How could inhaling smaller amounts of air and having more carbon dioxide in our bloodstream increase oxygen in our tissues and organs? How could doing less give us more?

To understand this contrarian concept, you need to consider the body parts beyond the nose and mouth. Those structures, after all, are simply the gateways for the long journey of breath. The purpose of the 25,000 inhales and exhales we take daily lies deeper inside us. And the farther we follow this air, the more surprising and strange the journey gets.

Your body, like all human bodies, is essentially a collection of tubes. There are wide tubes, like the throat and sinuses, and very thin tubes like capillaries. The tubes that make up the tissues of the lungs are very small, and we've got a lot of them. If you lined up all the tubes in the airways of your body, they'd reach from New York City to Key West—more than 1,500 miles.

Each breath you take must first travel down the throat, past a crossroads called the tracheal carina, which splits it into the right and left lungs. As it keeps going, that breath gets pushed into smaller tubes called the bronchioles until it dead-ends at 500 million little bulbs called the alveoli.

What happens next is complicated and confusing. An analogy may help.

Let's say you're about to take a river cruise. You're in a waiting room at the dock when a ship approaches. You pass through security, board the ship, and head off. This is similar to the path oxygen molecules take once they reach the alveoli. Each of these little "docking stations" is surrounded by a river of plasma filled with red blood cells. As these cells pass by, oxygen molecules will slip through the membranes of the alveoli and lodge themselves inside one.

The cellular cruise ship is filled with "guest rooms." In your blood cells, those rooms are the protein called hemoglobin. Oxygen takes a seat inside a hemoglobin; then the red blood cells journey upstream, deeper into the body.

As blood passes through tissues and muscles, oxygen will disembark, providing fuel to hungry cells. As oxygen offloads, other passengers, namely carbon dioxide—the "waste product" of metabolism—will pile aboard, and the cruise ship will begin a return journey back to the lungs.

Blood will grow darker as oxygen leaves. The blood in the veins will appear more bluish (it's actually darker red) because of the way in which light penetrates skin. Blue light has a shorter, stronger wavelength than other colors, which is also why the ocean and sky appear blue at a distance.

Eventually, the cruise ship will make its round through the body and back to port, back to the lungs, where carbon dioxide will exit the body through the alveoli, up the throat, and out the mouth and nose in an exhale. More oxygen boards in the next breath and the process starts again.

The cells in our bodies are fueled by oxygen, and this is how they get it. The entire cruise takes about a minute, and the overall numbers are staggering. Inside each of our 25 trillion red blood cells are 270 million hemoglobins, each of which has room for four oxygen molecules. That's *a billion molecules of oxygen* boarding and disembarking within each red blood cell cruise ship.

There's nothing controversial about this process of respiration and the role of carbon dioxide in gas exchange. It's basic biochemistry. What's less acknowledged is the role carbon dioxide plays in weight loss. That carbon dioxide in every exhale has weight, and we exhale more weight than we inhale. And the way the body loses weight isn't through profusely sweating or "burning it off." We lose weight through exhaled breath.

For every ten pounds of fat lost in our bodies, eight and a half pounds of it comes out through the lungs; most of it is carbon dioxide mixed with a bit of water vapor. The rest is sweated or urinated out. This is a fact that most doctors, nutritionists, and other medical professionals have historically gotten wrong. The lungs are the weight-regulating system of the body.

"Everyone always talks about oxygen," Olsson told me during our interview in Stockholm. "Whether we breathe thirty times or five times a minute, a healthy body will always have enough oxygen!"

What our bodies really want, what they require to function properly, isn't faster or deeper breaths. It's not more air. What we need is more carbon dioxide.

．　．　．

More than a century ago, a baggy-eyed Danish physiologist named Christian Bohr discovered this in a laboratory in Copenhagen. By his early 30s, Bohr had earned degrees in medicine and physiology and was working at the University of Copenhagen. He was fascinated with respiration; he knew that oxygen was the cellular fuel and that hemoglobin was the transporter. He knew that when oxygen went into a cell, carbon dioxide came out.

But Bohr didn't know *why* this exchange took place. Why did some cells get oxygen more easily than others? What directed billions of hemoglobin molecules to release oxygen at just the right place at the right time? How did breathing really work?

He began experimenting. Bohr gathered chickens, guinea pigs, grass snakes, dogs, and horses, and measured how much oxygen the animals consumed and how much carbon dioxide they produced. Then he drew blood and exposed it to different mixtures of these gases. Blood with the most carbon dioxide in it (more acidic) loosened oxygen from hemoglobin. In some ways, carbon dioxide worked as a kind of divorce lawyer, a go-between to separate oxygen from its ties so it could be free to land another mate.

This discovery explained why certain muscles used during exercise received more oxygen than lesser-used muscles. They were producing more carbon dioxide, which attracted more oxygen. It was supply on demand, at a molecular level. Carbon dioxide also had a profound dilating effect on blood vessels, opening these pathways so they could carry more oxygen-rich blood to hungry cells. Breathing less allowed animals to produce more energy, more efficiently.

Meanwhile, rapid and panicked breaths would purge carbon dioxide. Just a few moments of heavy breathing above metabolic needs could cause reduced blood flow to muscles, tissues, and organs. We'd feel lightheaded, cramp up, get a headache, or even black out. If these tissues were denied consistent blood flow for long enough, they'd break down.

In 1904, Bohr published a paper called "Concerning a Biologically Important Relationship—The Influence of the Carbon Dioxide Content of Blood on Its Oxygen Binding." It was a sensation among scientists and inspired a flurry of new research into this long-misunderstood gas. Soon after, Yandell Henderson, the director of the Laboratory of Applied Physiology at Yale, began his own set of experiments. Henderson had spent the last several years studying metabolism, and, like Bohr, he too was convinced that carbon dioxide was as essential to the body as any vitamin.

"Although clinicians still find it hard to believe, oxygen is in no sense a stimulant to living creatures," Henderson would write in the *Cyclopedia of Medicine*. "If a fire is supplied with pure oxygen instead of air, it burns with enormously augmented intensity. But when a man or animal breathes oxygen, or [air] enriched with oxygen, no more of that gas is consumed, no more heat is produced and no more carbon dioxide is exhaled than when air alone is breathed."

For a healthy body, overbreathing or inhaling pure oxygen would have no benefit, no effect on oxygen delivery to our tissues and organs, and could actually create a state of oxygen deficiency, leading to relative suffocation. In other words, the pure oxygen a quarterback might huff between plays, or that a jet-lagged traveler might shell out 50 dollars for at an airport "oxygen bar," are of no benefit. Inhaling the gas might increase blood oxygen levels one or two percent, but that oxygen will never make it into our hungry cells. We'll simply breathe it back out.*

To prove his point, over the years Henderson conducted a number of awful experiments on dogs that are about as difficult to read about as Harvold's awful experiments on monkeys.

He placed individual dogs on a table in his laboratory and inserted a tube into their throats, fitting their faces with a rubber mask. At the end of the tube was a hand bellows. The contraption allowed Henderson to control how much air each dog took in and how often. He'd connected the tube from the dogs' throats to a bottle of ether, which would anesthetize them during the course of the experiment. A suite of instruments recorded heart rate, carbon dioxide, oxygen levels, and more.

As Henderson pumped the bellows faster and faster, he watched heart rates of the animals quickly increase from 40 up to 200 or more beats per minute. The dogs would eventually have so much oxygen flowing through their arteries, with so little carbon dioxide to offload it, that

---

*Henderson discovered, one hundred years ago, that pure oxygen is only useful for those at altitude (where oxygen levels in the air are decreased) or those who are so sick that they cannot retain healthy oxygen saturation levels (above roughly 90 percent) through normal breathing. But even for sick patients, long-term supplemental oxygen can eventually damage the lungs and decrease red blood cell counts, making it harder for the body to pull oxygen from breath in the future.

muscles, tissues, and organs began failing. Some dogs would spasm uncontrollably or drift into a coma. If Henderson kept pumping more air in, the animals became so full of oxygen and so deficient in carbon dioxide that they died.

Henderson killed dogs with their own breath.

With the dogs that survived, he would pump the bellows slower and watch as their heart rates immediately decreased to 40 beats per minute. It wasn't the act of breathing that sped up and slowed the dogs' heart rates; it was the amount of carbon dioxide flowing through the bloodstream.

Henderson then forced the dogs to breathe *just slightly* harder than normal, just above their metabolic needs, so that their heart rates were mildly elevated and carbon dioxide levels a little deficient. This was a condition of mild hyperventilation common in humans.

The dogs grew agitated, confused, anxious, and glassy-eyed. The slight overbreathing was inducing the same confused state that occurred during altitude sickness or panic attacks. Henderson administered morphine and other drugs to slow the animals' heart rates closer to normal. The drugs worked partly because, as Henderson observed, they helped raise carbon dioxide levels.

But there was another way to restore the animals to health: let them breathe slowly. Whenever Henderson lowered the respiratory rate in accordance with the dogs' normal metabolism—from breathing 200 times a minute to a normal rate—all the twitching, stupors, and anxiety went away. The animals stretched out and relaxed, their muscles loosened, and a peacefulness washed over them.

"Carbon dioxide is the chief hormone of the entire body; it is the only one that is produced by every tissue and that probably acts on every organ," Henderson later wrote. "Carbon dioxide is, in fact, a more fundamental component of living matter than is oxygen."

. . .

I spent three days with Olsson in Stockholm. We pored over tables and graphs and talked about Bohr and Henderson and other storied pulmonauts. By the end of my trip, I finally understood how my own view of breathing had been so limited, and so wrong, for so many years. And I finally understood how Olsson had become so obsessed with this line of research, why he'd given up his life as a software tycoon and downgraded to a tiny apartment, surrounded by shelves of biochemistry textbooks, sleep tape, and carbon dioxide tanks. Why he'd spent so many months recording how carbon dioxide levels changed inside his body with each new breathing technique, how it affected his blood pressure and his energy and stress levels.

I understood why only one person showed up for the first conference he held on breathing, in 2010, and why, after honing his message and building his research base, he was now something of a Swedish media star who filled auditoriums, his grinning, perpetually tanned, rom-com face popping up on newspapers, magazines, and nightly news shows. In these interviews, he championed the therapeutic effects of nasal breathing and beseeched audiences with the same message of slow breathing.

I returned home to San Francisco, and Olsson and I kept in contact. Every few weeks I'd get a new email or a Skype call about some new long-lost scientific discovery he'd just unearthed in a medical library. He'd continued his self-experimentation too, always seeking to use his own body to prove the power of breathing and wonders of the "metabolic waste product," carbon dioxide.

This is how Olsson ended up, a year after our first meeting, in my living room in San Francisco with a face mask Velcroed to his head and an EKG electrode clipped to his ear.

. . .

"Could you hand me the oximeter, please?" Olsson says again across the table.

We've just finished our afternoon testing, and Olsson is restrapping himself into the BreathIQ, a prototype of a device that measures carbon dioxide, ammonia, and other elements in the exhaled breath. He clips a pulse oximeter on his finger and starts counting down the seconds.

Maybe it's the carbon dioxide and nitric oxide boost from nasal breathing, but we're feeling punchy today. In addition to the five grand we dropped to take before-and-after X-rays and blood and pulmonary function tests at Stanford, Olsson and I also managed to amass several thousand dollars' worth of equipment at the home lab. We've spent two weeks running tests and have yet to push the throttle on it all. That's changing today.

Olsson wipes a hand on his Abercrombie sweatshirt and scoots over so that I can see the readouts on the machines. All his vitals are normal: heart rate hovers around 75, systolic blood pressure clocks in at 126, oxygen levels at 97 percent. *Three, two, one,* he begins breathing.

But slowly, very slowly. He inhales and exhales three times slower than the average American, turning those 18 breaths a minute into six. As he sips air in through his nose and out through his mouth, I watch as his carbon dioxide levels rise from 5 percent to 6 percent. They keep rising. A minute later, Olsson's levels are 25 percent higher than they were just a few minutes ago, taking him from an unhealthy hypocapnic zone to squarely within a medically normal range. All the while, his blood pressure drops about five points and heart rate sinks to the mid-60s.

What hasn't changed is his oxygen. From start to finish, even though he's been breathing at a third of the rate considered normal, his oxygen hasn't wavered: it stayed at 97 percent.

We'd experienced the same confounding measurements during our bike workouts earlier in the week. The beginning of those workouts, like all workouts, sucked. We felt our lungs and respiratory system desperately trying to meet the needs of our hungry tissues and muscles: the dinner rush of the body. Normally, I'd open my mouth and huff and puff, trying to sate that nagging need for oxygen. But for the last few days, as I cranked the pedals harder and faster, I forced myself to breathe softer and slower. This felt stifling and claustrophobic, like I was starving my body of fuel, until I checked the pulse oximeter. Once again, no matter how slowly I breathed or how hard I pedaled, my oxygen levels held steady at 97 percent.

It turns out that when breathing at a normal rate, our lungs will absorb only about a quarter of the available oxygen in the air. The majority of that oxygen is exhaled back out. By taking longer breaths, we allow our lungs to soak up more in fewer breaths.

"If, with training and patience, you can perform the same exercise workload with only 14 breaths per minute instead of 47 using conventional techniques, what reason could there be not to do it?" wrote John Douillard, the trainer who'd conducted the stationary bike experiments in the 1990s. "When you see yourself running faster every day, with your breath rate stable . . . you will begin to feel the true meaning of the word *fitness.*"

I realized then that breathing was like rowing a boat: taking a zillion short and stilted strokes will get you where you're going, but they pale in comparison to the efficiency and speed of fewer, longer strokes.

On the second day of using this slower, nasal breathing approach, I'd outdistanced my mouthbreathing record by .13 of a mile. The next session, I pedaled .36 miles farther—a 5 percent increase over mouthbreathing. By my fifth ride on the stationary bike, I pedaled 7.7 miles, almost a full mile longer, in the same amount of time, using the same amount of energy, than I had the previous week. This was a significant gain. It wasn't quite yet up to the levels Douillard's cyclists reported, but I was edging closer.

During that ride, I started playing around with my breathing. I tried to inhale and exhale slower and slower, from my usual exercising rate of 20 breaths a minute to just six. I immediately felt a sense of air hunger and claustrophobia. After a minute or so I looked down at the pulse oximeter to see how much oxygen I was losing, how starved my body had become.

But my oxygen hadn't decreased with these very slow breaths, as I or anyone else might expect. My levels *rose*.

.   .   .

A last word on slow breathing. It goes by another name: prayer.

When Buddhist monks chant their most popular mantra, *Om Mani Padme Hum*, each spoken phrase lasts six seconds, with six seconds to inhale before the chant starts again. The traditional chant of *Om*, the "sacred sound of the universe" used in Jainism and other traditions, takes six seconds to sing, with a pause of about six seconds to inhale.

The *sa ta na ma* chant, one of the best-known techniques in Kundalini yoga, also takes six seconds to vocalize, followed by six seconds to inhale. Then there were the ancient Hindu hand and tongue poses called *mudras*. A technique called *khechari*, intended to help boost physical and spiritual health and overcome disease, involves placing the tongue above the soft palate so that it's pointed toward the nasal cavity. The deep, slow breaths taken during this *khechari* each take six seconds. Japanese, African, Hawaiian, Native American, Buddhist, Taoist, Muslim—these cultures and religions all had somehow developed the same prayer techniques, requiring the same breathing patterns. And they all likely benefited from the same calming effect.

In 2001, researchers at the University of Pavia in Italy gathered two dozen subjects, covered them with sensors to measure blood flow, heart rate, and nervous system feedback, then had them recite a Buddhist mantra as well as the original Latin version of the rosary, the Catholic prayer cycle of the Ave Maria, which is repeated half by a priest and half by the congregation. They were stunned to find that the average number of breaths for each cycle was "almost exactly" identical, just a bit quicker than the pace of the Hindu, Taoist, and Native American prayers: 5.5 breaths a minute.

But what was even more stunning was what breathing like this did to the subjects. Whenever they followed this slow breathing pattern, blood flow to the brain increased and the systems in the body entered a state of coherence, when the functions of heart, circulation, and nervous system are coordinated to peak efficiency. The moment the subjects returned to spontaneous breathing or talking, their hearts would beat a little more erratically, and the integration of these systems would slowly fall apart. A few more slow and relaxed breaths, and it would return again.

A decade after the Pavia tests, two renowned professors and doctors in New York, Patricia Gerbarg and Richard Brown, used the same breathing pattern on patients with anxiety and depression, minus the praying. Some of these patients had trouble breathing slowly, so Gerbarg and Brown recommended they start with an easier rhythm of three-second inhales with at least the same length exhale. As the patients got more comfortable, they breathed in and breathed out longer.

It turned out that the most efficient breathing rhythm occurred when both the length of respirations and total breaths per minute were locked in to a spooky symmetry: 5.5-second inhales followed by 5.5-second exhales, which works out almost exactly to 5.5 breaths a minute. This was the same pattern of the rosary.

The results were profound, even when practiced for just five to ten minutes a day. "I have seen patients transformed by adopting regular breathing practices," said Brown. He and Gerbarg even used this slow breathing technique to restore the lungs of 9/11 survivors who suffered from a chronic and painful cough caused by the debris, a horrendous condition called ground-glass lungs. There was no known cure for this ailment, and yet after just two months, patients achieved a significant improvement by simply learning to practice a few rounds of slow breathing a day.

Gerbarg and Brown would write books and publish several scientific articles about the restorative power of the slow breathing, which would become known as "resonant breathing" or Coherent Breathing. The technique required no real effort, time, or thoughtfulness. And we could do it anywhere, at any time. "It's totally private," wrote Gerbarg. "Nobody knows you're doing it."

In many ways, this resonant breathing offered the same benefits as meditation for people who didn't want to meditate. Or yoga for people who didn't like to get off the couch. It offered the healing touch of prayer for people who weren't religious.

Did it matter if we breathed at a rate of six or five seconds, or were a half second off? It did not, as long as the breaths were in the range of 5.5.

"We believe that the rosary may have partly evolved because it synchronized with the inherent cardiovascular (Mayer) rhythms, and thus gave a feeling of wellbeing, and perhaps an increased responsiveness to the religious message," the Pavia researchers wrote. In other words, the meditations, Ave Marias, and dozens of other prayers that had been developed over the past several thousand years weren't all baseless.

Prayer heals, especially when it's practiced at 5.5 breaths a minute.

Six

LESS

Few would dispute that we've become a culture of overeaters. From around 1850 to 1960, the American mean body mass index (BMI), a measurement of fat based on height, was between 20 and 22. That's about 160 pounds for a six-foot-tall person. Today, the average BMI is 29, a 38 percent jump in 50 years. That six-foot person now weighs 214 pounds. Seventy percent of the U.S. population is considered overweight; one in three are obese. There's no doubt we are eating more than we did in the past.

Rates of breathing are much more difficult to gauge, because there are fewer studies and the results are inconsistent. Nonetheless, a review of several available studies offers a troubling picture.

What's considered medically normal today is anywhere between a dozen and 20 breaths a minute, with an average intake of about half a liter per breath. For those on the high end of respiratory rates, that's about twice as much as it was.*

One thing that every medical or freelance pulmonaut I've talked to over the past several years has agreed on is that, just as we've become a culture of overeaters, we've also become a culture of overbreathers. Most of us breathe too much, and up to a quarter of the modern population suffers from more serious chronic overbreathing.

The fix is easy: breathe less. But that's harder than it sounds. We've become conditioned to breathe too much, just as we've been conditioned to eat too much. With some effort and training, however, breathing less can become an unconscious habit.

Indian yogis train themselves to decrease the amount of air they take in at rest, not increase it. Tibetan Buddhists prescribed step-by-step instructions to reduce and calm breathing for aspiring monks. Chinese doctors two thousand years ago advised 13,500 breaths per day, which works out to nine and a half breaths per minute. They likely breathed less in those fewer breaths. In Japan, legend has it that samurai would test a soldier's readiness by placing a feather beneath his nostrils while he inhaled and exhaled. If the feather moved, the soldier would be dismissed.

To be clear, breathing less is not the same as breathing slowly. Average adult lungs can hold about four to six liters of air. Which means that, even if we practice slow breathing at 5.5 breaths per minute, we could still be easily taking in twice the air we need.

The key to optimum breathing, and all the health, endurance, and longevity benefits that come with it, is to practice fewer inhales and exhales in a smaller volume. To breathe, but to *breathe less*.

---

*See studies and other references in Notes, p. 246, at "offers a troubling picture."

. . .

With only four days left in the Stanford experiment, I was reaping the benefits of slowing my respiratory rate. My blood pressure kept dropping, my heart rate variability kept rising, and I had more energy than I knew what to do with.

All the while, Olsson kept prodding me to reduce my breathing rate even further. He harped on about the wonders of breathing *way less* than anyone normally should: the respiratory equivalent of fasting. Starving yourself of air can be injurious if it becomes a regular thing, he warned. Ordinarily, we should breathe as closely in line with our needs as we can. But occasionally willing the body to breathe *way less*, he argued, has some potent benefits just as fasting does. Sometimes it can lead to euphoria.

"It was a better feeling than when I got married, better than when my first kid was born," Olsson says.

It's morning and we're driving past the ragged gray waves along Highway 1. I'm at the wheel and Olsson is beside me in the passenger seat, smiling broadly, reliving the moment five years ago when he saw God.

"I ran for an hour or so, six miles I think, and I came home and sat in my living room chair." His voice is quivering a little here, he's almost laughing. "And I had this dull headache, that *good* headache, and I felt the most intense peace and unity in the world . . . everything. . . ."

Our destination today is Golden Gate Park, which offers miles of uninterrupted jogging tracks beneath the canopies of blue gum eucalyptus, Tasmanian tree ferns, cypress, and redwoods. Because the tracks are dirt, we won't split our heads open and die if we suddenly go unconscious, which, Olsson warns, is a rare but real side effect of the breathing-*way-less* thing we'll be attempting.

Olsson swears by this approach. He and his clients have reported profound improvements in endurance and well-being after a few weeks of training. However, I'd heard from many others that it could be miserable and might induce vicious headaches—not "good" ones. It wasn't for dabblers.

I turn the car from the freeway on to a one-lane street and park beside the lot of the Golden Gate Angling and Casting Club. A herd of buffalo behind a chain-link fence stare with bored eyes as Olsson and I slip off our jackets, take a few last swigs of water, lock the car, and hit the ground running.

I hate jogging. Unlike with other physical activities—especially water sports like surfing or swimming—whenever I jog, I'm fully conscious of the misery and boredom of every second. I've never reached that spaced-out runner's high, even though, years back, I'd put in four-mile runs every other day. The benefits of jogging were obvious: I always felt great . . . afterward. But doing it was a grind.

Olsson wanted to change my mind. He's been jogging for decades and has trained dozens of runners. "The key is to find a rhythm that works for you," he tells me as we beeline into the bramble. "You should challenge yourself, but at the same time don't overdo it."

The trail splits off and we follow the path less traveled. The sun shines through skyscraper trees, there's musty spearmint wafting through the air and the satisfying crunch of footfalls on crispy leaves. It's nice.

"What I want you to do is, as you warm up, start extending your exhales," he says. He prepped me earlier for this, so I know what's coming.

Each breath we draw in should take about three seconds, and each breath out should take four. We'll then continue the same short inhales while lengthening the exhales to a five, six, and seven count as the run progresses.

Slower, longer exhales, of course, mean higher carbon dioxide levels. With that bonus carbon dioxide, we gain a higher aerobic endurance. This measurement of highest oxygen consumption, called VO_2 max, is the best gauge of cardiorespiratory fitness. Training the body to breathe less actually increases VO_2 max, which can not only boost athletic stamina but also help us live longer and healthier lives.

. . .

The godfather of less-is-more was a pulmonaut born in 1923 on a farm outside Kiev, in what is now Ukraine. His name was Konstantin Pavlovich Buteyko, and he spent his youth examining the world around him. Anything, really. Plants, insects, toys, cars. He came to view the world as a mechanism, and everything within it as a collection of parts locking together to form a greater whole. By the time he was a teenager, Buteyko had developed into a brilliant mechanic and would later spend four years on the front lines of World War II fixing cars, tanks, and artillery for the Soviet army.

"When the War ended, I decided to start researching the most complex machine, the Man," he said. "I thought if I learnt him, I'd be able to diagnose his diseases as easily as I had diagnosed machine disorders."

Buteyko went on to attend the First Moscow Institute of Medicine, the most prestigious medical school in the Soviet Union, graduating cum laude in 1952. During his residency rounds, he noticed that patients in the worst health all seemed to breathe far too much. The more they breathed, the worse off they were, especially those with hypertension.

Buteyko himself suffered from severe high blood pressure, along with the debilitating headaches and stomach and heart pain that often accompanied the condition. He'd been put on prescription drugs to no effect. When he was 29, his systolic blood pressure had shot to 212, a dangerously high number. Doctors gave him a year to live.

"One can avoid cancer by cutting it out," Buteyko would later say. "But you can't avoid hypertension." The best he could do for his patients, and himself, was to try to numb the symptoms.

As the story goes, one night in October, Buteyko was standing alone in a hospital room, looking out from a window into a black autumn sky. He turned his focus to his reflection in the glass—a gaunt and haggard face drawing heavy breaths through an open mouth. His eyes roamed down to the white robe covering his chest, to his shoulders flexing and lifting with each labored inhale and exhale. This was the same respiratory rate he'd seen in terminally ill patients. Buteyko wasn't exercising, and yet he was breathing as if he'd just finished a workout.

He tried an experiment. He started breathing less, to relax his chest and stomach and sip air through his nose. A few minutes later, the throbbing pain in his head, stomach, and heart disappeared. Buteyko returned to the heavy breathing he'd been doing minutes earlier. Within only five inhales, the pain returned.

What if overbreathing wasn't the result of hypertension and headaches but the cause? Buteyko wondered. Heart disease, ulcers, and chronic inflammation were all linked to disturbances in circulation, blood pH, and metabolism. How we breathe affects all those functions. Breathing just 20 percent, or even 10 percent more than the body's needs could overwork our systems. Eventually, they'd weaken and falter. Was breathing too much making people sick, and keeping them that way?

Buteyko took a walk. In the asthma ward, he found a man stooped over, fighting suffocation, gasping for air. Buteyko approached and showed him the technique he'd been using on himself. After a few minutes, the patient calmed down. He inhaled a careful and clear breath through his nose and then calmly exhaled. Suddenly, his face flushed with color. The asthma attack was over.

. . .

Back in Golden Gate Park, Olsson and I are jogging deeper into the foot trail. The bucolic scene of dappled sunlight and *Avatar* trees has morphed into a more urban mess of wheel-less shopping carts and suspicious mounds of toilet paper. We realize the path less traveled may be less traveled for a reason. A quick left, and we're on our way back to the coastal route.

We jog past an old hippie sitting on a tree stump playing the *Jeopardy!* theme song on a trumpet with one hand and reading a dog-eared paperback with the other. In front of him, an impeccably dressed man goads an old dog into a beat-up Mercedes 300SD, and a woman with waist-length dreadlocks and Mork-from-Ork suspenders whizzes by on an electric scooter. It's a quintessentially San Francisco scene. Olsson and I fit right in.

We've been practicing an extreme version of the techniques Buteyko used on himself and in the asthma ward: limiting our inhales while extending exhales far past the point of what feels comfortable, or even safe. We're sweating and red-faced and I can feel veins bulging in my neck. I'm not exactly out of breath, but I don't feel satisfied, either. Even when I sip a little more air, I feel like I'm being mildly strangled.

The point of this exercise isn't to inflict unnecessary pain. It's to get the body comfortable with higher levels of carbon dioxide, so that we'll unconsciously breathe less during our resting hours and the next time we work out. So that we'll release more oxygen, increase our endurance, and better support all the functions in our bodies.

"Try to extend the exhales even more," Olsson says as he takes tiny nips of air through his nose. "Breathe out twice as long for each inhale, three times," he chides me. For a moment, I feel like I'm going to puke.

"Yes!" he says. "Even slower, even less!"

. . .

By the late 1950s, Buteyko left the hospitals of Moscow and headed to Akademgorodok ("Academic City"), a cluster of 35 concrete-block research facilities located in central Siberia. The distant location was by design. For the past few years, the Soviet government had sent tens of thousands of the finest space engineers, chemists, physicists, and others to live in secrecy among the laboratories. Their job was to develop cutting-edge technologies aimed at ensuring the Soviet Union's dominance. In many ways it was a Soviet Silicon Valley, but without the fleece vests, kombucha, sunshine, Teslas, and civil liberties.

Buteyko had moved there at the request of the USSR Academy of Medical Sciences, the Soviet equivalent to the Centers for Disease Control and Prevention. After his epiphany in the asthma ward, he'd pored over research papers and analyzed hundreds of patients. He'd become convinced that breathing too much was the culprit behind several chronic diseases. Like Bohr and Henderson, Buteyko was fascinated with carbon dioxide, and he too believed that increasing this gas by breathing less could not only keep us fit and healthy. It could heal us as well.

At Akademgorodok he set out to conduct the most exhaustive breathing experiments that science had ever attempted. He collected a staff of more than 200 researchers and assistants in a sweeping city hospital called the Laboratory of Functional Diagnostics. Subjects would come in and lie on a gurney, sandwiched between stacks of machines. Phlebotomists would plug catheters into their veins while other researchers stuck hoses into their throats and electrodes around the heart and head. As the subjects breathed in and out, a primitive computer recorded 100,000 bits of data per hour.

The sick and healthy, young and old—more than a thousand of them came to Buteyko's lab. The patients with asthma, hypertension, and other ailments consistently breathed the same: too much. They often inhaled and exhaled through the mouth, packing in 15 liters or more of air per minute. Some breathed so loudly they could be heard several feet away. The readouts showed that they had plenty of oxygen in their blood, but much less carbon dioxide, about 4 percent. Resting heart rates were up to 90 beats per minute.

The healthiest patients breathed alike, too: less. They'd inhale and exhale about ten times a minute, taking in a total of about five to six liters of air. Their resting pulses ranged from around 48 to 55, and they had about 50 percent more carbon dioxide in their exhaled breath.

Buteyko developed a protocol based on the breathing habits of these healthiest patients, which he'd later call Voluntary Elimination of Deep Breathing. The techniques were many and they varied, but the purpose of each was to train patients to always breathe as closely as possible to their metabolic needs, which almost always meant taking in less air. How many breaths we took per minute was less important to Buteyko, as long as we were breathing no more than about six liters per minute at rest.

Within a few sessions of practicing these techniques, patients reported tingling and heat in their hands and toes. Their heart rates would slow and stabilize. The hypertension and migraines that had debilitated so many of them would begin to disappear. Those already in good health felt even better. Athletes claimed big gains in performance.

Around this time, a few thousand miles west, in the industrial factory town of Zlín, Czechoslovakia, a gangly, five-foot-eight-inch runner named Emil Zátopek was experimenting with his own breath-restriction techniques.

Zátopek never wanted to become a runner. When the management at the shoe factory where he was working elected him for a local race, he tried to refuse. Zátopek told them he was unfit, that he had no interest, that he'd never run in a competition. But he competed anyway and came in second out of 100 contestants. Zátopek saw a brighter future for himself in running, and began to take the sport more seriously. Four years later he broke the Czech national records for the 2,000, 3,000, and 5,000 meters.

Zátopek developed his own training methods to give himself an edge. He'd run as fast as he could holding his breath, take a few huffs and puffs and then do it all again. It was an extreme version of Buteyko's methods, but Zátopek didn't call it Voluntary Elimination of Deep Breathing. Nobody did. It would become known as hypoventilation training. *Hypo*, which comes from the Greek for "under" (as in hypodermic needle), is the opposite of *hyper*, meaning "over." The concept of hypoventilation training was to breathe less.

Over the years, Zátopek's approach was widely derided and mocked, but he ignored the critics. At the 1952 Olympics, he won gold in the 5,000 and 10,000 meters. On the heels of his success, he decided to compete in the marathon, an event he had neither trained for nor run in his life. He won gold. Zátopek would claim 18 world records, four Olympic golds and a silver over his career. He would later be named the "Greatest Runner of All Time" by *Runner's World* magazine. "He does everything wrong but win," said Larry Snyder, a track coach at Ohio State at the time.

Hypoventilation training didn't exactly take off after Zátopek. His anguished face, teeth grinding and eyes wincing like a Matthias Grünewald *Jesus*, became his trademark look as he crossed the finish line, often in first place. It all seemed miserable, because it was, and most athletes steered clear.

Then, decades later, in the 1970s, a hard-assed U.S. swim coach named James Counsilman rediscovered it. Counsilman was notorious for his "hurt, pain, and agony"–based training techniques, and hypoventilation fit right in.

Competitive swimmers usually take two or three strokes before they flip their heads to the side and inhale. Counsilman trained his team to hold their breath for as many as nine strokes. He believed that, over time, the swimmers would utilize oxygen more efficiently and swim faster. In a sense, it was Buteyko's Voluntary Elimination of Deep Breathing and Zátopek hypoventilation—underwater. Counsilman used it to train the U.S. Men's Swimming team for the Montreal Olympics. They won 13 gold medals, 14 silver, and 7 bronze, and they set world records in 11 events. It was the greatest performance by a U.S. Olympic swim team in history.

Hypoventilation training fell back into obscurity after several studies in the 1980s and 1990s argued that it had little to no impact on performance and endurance. Whatever these athletes were gaining, the researchers reported, must have been based on a strong placebo effect.

In the early 2000s, Dr. Xavier Woorons, a French physiologist at Paris 13 University, found a flaw in these studies. The scientists critical of the technique had measured it all wrong. They'd been looking at athletes holding their breath with full lungs, and all that extra air in the lungs made it difficult for the athletes to enter into a deep state of hypoventilation.

Woorons repeated the tests, but this time subjects practiced the half-full technique, which is how Buteyko trained his patients, and likely how Counsilman trained his swimmers. Breathing less offered huge benefits. If athletes kept at it for several weeks, their muscles adapted to tolerate more lactate accumulation, which allowed their bodies to pull more energy during states of heavy anaerobic stress, and, as a result, train harder and longer. Other reports showed hypoventilation training provided a boost in red blood cells, allowing athletes to carry more oxygen and produce more energy with each breath. Breathing *way less* delivered the benefits of high-altitude training at 6,500 feet, but it could be used at sea level, or anywhere.

Over the years, this style of breath restriction has been given many names—hypoventilation, hypoxic training, Buteyko technique, and the pointlessly technical "normobaric hypoxia training." The outcomes were the same: a profound boost in performance.* Not just for elite athletes, but for everyone.

Just a few weeks of the training significantly increased endurance, reduced more "trunk fat," improved cardiovascular function, and boosted muscle mass compared to normal-breathing exercise. This list goes on.

The takeaway is that hypoventilation works. It helps train the body to do more with less. But that doesn't mean it's pleasant.

· · ·

Olsson and I emerge from the shady tranquility of Golden Gate Park, stopping to face the wind-ripped Pacific Ocean. We've just jogged a few miles, inhaling fast and exhaling very long breaths to a count of about seven or higher, trying to keep our lungs roughly half full. I want to believe that this training may be helping me as it helped Zátopek, Counsilman's swimmers, Wooron's runners, and everyone else, but the past several minutes have been a challenge. A half hour into all this I'm beginning to resent my life choices. I can't figure out if it's bad luck or shortsightedness that has led me to repeatedly pursue topics of research like freediving, Voluntary Elimination of Deep Breathing, and hypoventilation therapy that require me to hold my breath and torture my lungs for hours a day.

"The key is to find a rhythm that works for you," Olsson keeps saying. The rhythm is definitely *not* working. I return to my more manageable practice, inhaling for two steps and exhaling for five, a pattern competitive cyclists use. This isn't exactly comfortable, but it's tolerable.

We run across the cracked asphalt of a beachside parking lot, passing a few rusting Winnebagos and hopping over condom wrappers and smashed

---

*More recently, Sanya Richards-Ross, a Jamaican-American sprinter, used Buteyko's techniques to win three Olympic golds in the 4x400 meter relay (in 2004, 2008, and 2012) and gold in the 400 meters in 2012. She was ranked as the top 400-meter runner in the world for a decade. Photos of Richards-Ross with her mouth closed and a placid look on her face as she destroys the slack-jawed and gasping competition have become the stuff of legend.

cans of malt liquor before heading back across the highway. Minutes later we're back in the quietude of the park, treading a dirt path beneath an understory of trees along a black pond filled with quacking ducks.

That's when it starts to hit me: an intense heat at the back of my neck and pixelated vision. I'm still jogging, exhaling long breaths, but it feels as though I'm simultaneously jumping headfirst into warm, thick liquid. I run a little harder, breathe a little less, and feel heat, heavy like hot syrup, seeping down into my fingertips, toes, arms, and legs. It feels great. The warmth moves higher through my face and wraps around the crown of my head.

This must be the *good headache* Olsson was talking about, of carbon dioxide increasing and oxygen dislodging from hemoglobin to those hungry cells, of the vessels in my brain and body expanding, so engorged with fresh blood that they're sending dull pain signals to my nervous system.

Just when it feels like I'm about to reach some kind of existential crescendo, the little foot trail widens. The bored buffalo appear, rustling behind a chain-link fence. A dozen yards away is the lot of the Golden Gate Angling and Casting Club. My car is beside it, and we're done.

There are no huge life epiphanies as we drive home. I can't say I'm feeling euphoric, but that's OK. My little jog proved there's much to gain from this *less* approach. At the same time, such extreme training would be useful only for those willing to endure hours of red-faced, sweaty suffering.

Healthy breathing shouldn't be so much work. Buteyko knew this, and he rarely, if ever, prescribed such brutal methods to his patients. After all, he wasn't interested in coaching elite athletes to win gold medals. He wanted to save lives. He wanted to teach techniques of breathing less that could be practiced by everyone, regardless of their state of health, age, or level of fitness.

Over his career, Buteyko would be censured by medical critics; he'd be physically attacked and, at one point, have his laboratory torn up. But he pressed on. By the 1980s, he had published more than 50 scientific papers and the Soviet Ministry of Health had recognized his techniques as effective. Some 200,000 people in Russia alone had learned his methods. According to several sources, Buteyko was once invited to England to meet with Prince Charles, who was suffering from breathing difficulties brought on by allergies. Buteyko helped the prince, and he helped heal upward of 80 percent of his patients suffering from hypertension, arthritis, and other ailments.

Voluntary Elimination of Deep Breathing was especially effective in treating respiratory diseases. It seemed to work like a miracle for asthma.

.   .   .

In the decades since Buteyko first started training patients to breathe less, asthma has become a global epidemic. Nearly 25 million Americans now suffer from it—that's about 8 percent of the population, and a fourfold increase since 1980. Asthma is the leading cause of emergency room visits, hospitalizations, and missed school days for children. It is considered a controllable but incurable disease.

Asthma is an immune system sensitivity that provokes constriction and spasms in the airways. Pollutants, dust, viral infections, cold air, and more can all lead to attacks. But asthma can be brought on by overbreathing, which is why it's so common during physical exertion, a condition called exercise-induced asthma that affects around 15 percent of the population and up to 40 percent of athletes. At rest or during exercise, asthmatics as a whole tend to breathe more—sometimes much more—than those without asthma. Once an attack starts, things go from bad to worse. Air gets trapped in the lungs and passageways constrict, which makes it harder to push air out and back in. More breathing but more feelings of breathlessness ensue, more constriction, more panic, and more stress.

The worldwide annual market for asthma therapies is $20 billion, and drugs often work so well that they can feel like a virtual cure. But drugs, in particular oral steroids, can have horrendous side effects after several years, including deteriorating lung function, worsened asthma symptoms, blindness, and increased risk of death. Millions of asthma sufferers already know this, and are experiencing these problems for themselves. Many of them have trained themselves to breathe less and reported dramatic improvement.

For several months before the Stanford experiment, I interviewed Buteyko practitioners and collected their stories.

One was David Wiebe, a 58-year-old luthier of cellos and violins from Woodstock, New York, whom I'd read about in *The New York Times*. Wiebe had suffered from severe asthma since he was ten. He used bronchodilators up to 20 times a day, along with steroids, in an effort to keep his symptoms at bay. His body became tolerant of drugs, which meant that Wiebe had to increase the dose. After decades of constant use, the steroids weakened his eyesight, a condition called macular degeneration. If he kept taking them, Wiebe would go blind; if he stopped taking them, he wouldn't be able to breathe and might die of an asthma attack.

Within three months of learning how to breathe less, Wiebe was using no more than one inhaler puff a day, and he'd cut out steroids entirely. He claimed to feel few asthma symptoms at all. For the first time in five decades he could breathe easy. Even Wiebe's pulmonologist was impressed, confirming that there was a marked improvement in Wiebe's asthma and overall health.

There were others. Like the chief information officer at the University of Illinois at Urbana-Champaign, who had also suffered from debilitating asthma his whole adult life, and who, like Wiebe, reported few symptoms of asthma within weeks of retraining himself to breathe less. "I'm a new man," he wrote. There was the 70-year-old woman I'd spent an hour with at a Whole Foods café, who had experienced crippling asthma the past six decades and could hardly walk a few blocks without having an attack. After a few months of breathing less she was hiking for hours a day and on her way to travel in Mexico. "This is nothing short of a miracle," she told me. There was the mom from Kentucky who endured such horrendous breathing problems that she contemplated suicide. There were also athletes like Olympians Ramon Andersson, Matthew Dunn, and Sanya Richards-Ross, who had used breathing less methods. All of them claimed to have gained a boost in performance and blunted the symptoms of respiratory problems, simply by decreasing the volume of air in their lungs and increasing the carbon dioxide in their bodies.

The most convincing scientific validation of breathing less for asthma came by way of Dr. Alicia Meuret, director of the Anxiety and Depression Research Center at Southern Methodist University in Dallas. In 2014, Meuret and a team of researchers gathered 120 randomly selected asthma sufferers, measured their pulmonary lung functions, lung size, and blood gases, and then gave them a handheld capnometer, which tracked the carbon dioxide in their exhaled breath.

Over four weeks, the asthmatics would carry the device around and practice breathing less to keep their carbon dioxide levels at a healthy level of 5.5 percent. If the levels dipped, the patients would breathe less until the carbon dioxide levels rose back. A month later, 80 percent of the asthmatics had raised their resting carbon dioxide level and experienced significantly fewer asthma attacks, better lung function, and a widening of their airways. They all breathed better. Asthma symptoms were either gone or markedly decreased.

"When people hyperventilate, there is something very strange happening," Meuret wrote. "In essence they are taking in too much air. But the sensation that they get is shortness of breath, choking, air hunger, as if they're not getting enough air. It's almost like a biological system error." Willing the body to breathe less air appeared to correct that system error.

. . .

By the end of his career, and the end of his life in 2003 at the age of 80, Buteyko would become a bit of a mystic. He barely slept and claimed that his techniques could not only heal illnesses but promote intuition and other forms of extrasensory perception. He was convinced that heart disease, hemorrhoids, gout, cancer, and more than 100 other diseases were all caused by carbon dioxide deficiency brought on by overbreathing. He even considered asthma attacks less a problem, less a "system malfunction," and more a compensatory action. That airway constriction, wheezing, and shortness of breath was the body's natural reflex to breathe less and more slowly.

For these reasons and others, Buteyko and his methods have been largely dismissed by today's medical community as pseudoscience. Nonetheless, a few dozen researchers over the past few decades have attempted to gain some kind of real scientific validation on the restorative effects of breathing less. One study at the Mater Hospital in Brisbane, Australia, found that when asthmatic adults followed Buteyko's methods and decreased their air intake by a third, symptoms of breathlessness reduced by 70 percent and the need for reliever medication decreased by around 90 percent. A half-dozen other clinical trials showed similar results. Meanwhile, the Papworth Method, a breathing-less technique developed in an English hospital in the 1960s, was also shown to cut asthma symptoms by a third.*

---

*The primary complaint is that studies in Buteyko breathing were small, few, and according to some critics, done outside stringent scientific protocol. Be that as it may, in 2014, the Global Initiative for Asthma, a collaboration among the World Health Organization, the National Heart, Lung, and Blood Institute, and the National Institutes of Health, gave Buteyko an "A" rating (later revised to a "B") for supporting evidence.

Still, nobody seems to know exactly *why* breathing less has been so effective in treating asthma and other respiratory conditions. Nobody knows exactly how it works. There are several theories.

"It is a deficiency in the body that causes symptoms," said Dr. Ira Packman, an internist and former medical expert for the Pennsylvania Insurance Department who overcame his own debilitating asthma by breathing less. "Replace the deficient element," he told me, "and the patient gets better."

Packman explained that overbreathing can have other, deeper effects on the body beyond just lung function and constricted airways. When we breathe too much, we expel too much carbon dioxide, and our blood pH rises to become more alkaline; when we breathe slower and hold in more carbon dioxide, pH lowers and blood becomes more acidic. Almost all cellular functions in the body take place at a blood pH of around 7.4, our sweet spot between alkaline and acid.

When we stray from that, the body will do whatever it can to get us back there. The first response is from our cells. Sensing the pH is too alkalotic, cells secrete hydrogen ions that combine with bicarbonate to make the blood more acidic. This process is called cell buffering,* and it occurs about ten minutes into hyperventilation. Should we continue to overbreathe past two to six hours, the kidneys begin their own buffering by dumping bicarbonate into the urine, which further increases blood acidity.

The problem with buffering is that it's meant as a temporary fix, not a permanent solution. Weeks, months, or years of overbreathing, and this constant kidney (renal) buffering will deplete the body of essential

---

*Cells can also produce energy (ATP) anaerobically, without oxygen, and in doing so they create a more acidic "microenvironment" in which oxygen can more easily dissociate from hemoglobin. In this, chronic overbreathing will not create "hypoxia" in tissues; this is a fact that many Buteyko adherents consistently get wrong. The real damage from overbreathing comes from the constant energy the body has to expend to run more cells anaerobically and to constantly buffer for carbon dioxide deficiencies.

minerals. This occurs because as bicarbonate leaves the body through urine, it takes magnesium, phosphorus, potassium, and more with it. Without healthy stores of these minerals, nothing works right: nerves malfunction, smooth muscles spasm, and cells can't efficiently create energy. Breathing becomes even more difficult. This is one reason why asthmatics and other people with chronic respiratory problems are sometimes prescribed supplements like magnesium to stave off further attacks.

Constant buffering also weakens the bones, which try to compensate by dissolving their mineral stores back into the bloodstream. (Yes, it's possible to overbreathe yourself into osteoporosis and increased risk of bone fractures.) This unending grind of imbalances and compensations, of deficiencies and strain, will eventually break the body down.

Packman was quick to point out that not all respiratory illness sufferers and other sick people have a carbon dioxide deficiency problem. Those with emphysema, for instance, may have dangerously high levels of carbon dioxide because they've got too much stale air trapped inside. Others may test with completely normal blood gas and pH levels. But such nitpicking, he said, missed the larger point.

All these people have a *breathing problem*. They're stressed, they're inflamed, they're congested, and they struggle to get air in and out of their lungs. And it's these breathing problems that slow, paced, less techniques are so effective at fixing.

· · ·

Over several months leading up to the Stanford experiment, I visited with several Buteyko teachers and other low-breathing devotees. They told me the same story, of how they'd been plagued by some chronic respiratory illness that no drug or surgery or medical therapy could fix. Of how they all "cured" themselves with nothing more than breathing less. The techniques they used varied, but all circled around the same premise: to extend the length of time between inhalations and exhalations. The less one breathes, the more one absorbs the warming touch of respiratory efficiency—and the further a body can go.

This shouldn't come as much of a surprise. Nature functions in orders of magnitude. Mammals with the lowest resting heart rates live the longest. And it's no coincidence that these are consistently the same mammals that breathe the slowest. The only way to retain a slow resting heart rate is with slow breaths. This is as true for baboons and bison as it is for blue whales and us.

"The yogi's life is not measured by the number of his days, but the number of his breaths," wrote B. K. S. Iyengar, an Indian yoga teacher who had spent years in bed as a sickly child until he learned yoga and breathed himself back to health. He died in 2014, at age 95.

I'd hear this repeated over and over again by Olsson during our early Skype chats and again throughout the Stanford experiment. I'd read about it in Stough's research. Buteyko and the Catholics, Buddhists, Hindus, and 9/11 survivors were aware of it as well. By various means, in various ways, in various eras of human history, all these pulmonauts discovered the same thing. They discovered that the optimum amount of air we should take in at rest per minute is 5.5 liters. The optimum breathing rate is about 5.5 breaths per minute. That's 5.5-second inhales and 5.5-second exhales. This is the perfect breath.

Asthmatics, emphysemics, Olympians, and almost anyone, anywhere, can benefit from breathing this way for even a few minutes a day, much longer if possible: to inhale and exhale in a way that feeds our bodies just the right amount of air, at just the right time, to perform at peak capacity.

To just keep breathing, *less*.

Seven

CHEW

○———○

It's the nineteenth day of the Stanford experiment, and Olsson and I are, once again, sitting side by side at the dining room table at the center of our home lab. The place is officially a sty. We've just stopped caring. Because now we're just a few hours away from it all being over.

I'm sitting with the same thermometer and nitric oxide sensor in my mouth, and the same blood pressure cuff around my biceps. Olsson's got the same face mask around his head, the same EKG sensor on his ear. He's wearing the same slippers as well.

We've done this drill 60 times in the past three weeks. It all would have been unbearable had it not been for the mounting energy, mental clarity, and overall well-being we've felt, such vast and sudden improvements experienced the minute we stopped mouthbreathing.

Last night, Olsson snored for three minutes while I clocked in at six, a 4,000 percent decrease from ten days ago. Our sleep apnea, which disappeared the first night of nasal breathing, has remained nonexistent. My blood pressure this morning was 20 points lower than its highest point at the beginning of the experiment; on average I've dropped 10 points. My carbon dioxide levels consistently rose and were finally nudging toward the "super endurance" mark shared by Buteyko's healthiest subjects. Olsson, again, showed similar improvements. We did it all by breathing through our noses, slowly and less with full exhales.

"I'm done," declares Olsson, with the same smirk on his face. He walks down the hallway one last time and goes back across the street. And one last time, I'm left alone in the clutter, where I eat the same dinner I had ten days ago.

The last supper: a bowl of pasta, leftover spinach, a few soggy croutons. I take a seat at the kitchen table in front of the same unread pile of the Sunday *New York Times*, pour a little olive oil and salt into the bowl, and take a bite. A few mashes in my mouth and it's gone.

. . .

As random as it might seem, this mundane act—that few seconds of soft chewing—was the catalyst for writing this book. It's what inspired me to turn the casual hobby of investigating what happened to me in that Victorian room a decade ago into a full-time quest to discover the lost art and science of breathing.

At the beginning of this book, I began to fill in why humans have such a hard time breathing, how all that tenderizing and cooking of food eventually led to airway obstruction. But the changes that occurred in our heads and airways so many years ago were only a small part of how we got to here. There is a much deeper history to our origins, which is weirder and wilder than anything I had anticipated when I first began.

And so, here, at the end of the Stanford experiment, it seems only appropriate to start over again, picking up where we left off, at the dawn of human civilization.

. . .

Twelve thousand years ago, humans in Southwest Asia and the Fertile Crescent in the Eastern Mediterranean stopped gathering wild roots and vegetables and hunting game, as they had for hundreds of thousands of years. They started growing their food. These were the first farming cultures, and in these primitive communities, humans suffered from the first widespread instances of crooked teeth and deformed mouths.

It wasn't terrible at first. While one farming culture was plagued by facial and mouth deformities, another hundreds of miles away seemed not to suffer at all. Crooked teeth and all the breathing problems that come with them seemed totally random.

Then, about 300 years ago, these maladies went viral. Suddenly, all at once, much of the world's population began to suffer. Their mouths shrank, faces grew flatter, and sinuses plugged.

The morphological changes to the human head that had occurred up to this point—that lowering of the larynx that clogged our throats, the expansion of our brains that lengthened our faces—all these were negligible compared to this sudden shift. Our ancestors adapted to those gradual changes just fine.

But the changes triggered by the rapid industrialization of farmed foods were severely damaging. Within just a few generations of eating this stuff, modern humans became the worst breathers in *Homo* history, the worst breathers in the animal kingdom.

I had a hard time comprehending this when I first came across it years back. Why hadn't I been told about it in school? Why didn't so many of the sleep doctors or dentists or pulmonologists I interviewed know this story?

Because this research, I discovered, wasn't happening in the halls of medicine. It was happening in ancient burial sites. The anthropologists working at these sites told me that if I wanted to really understand how such a sudden and dramatic change could happen to us, and why, I needed to leave the labs and get into the field. I needed to see some of the Patient Zeros of the modern obstructed human, the turning point when our farm-fed faces fell apart on a mass scale. I needed to get my hands on some skulls: old ones, and lots of them.

I hadn't been introduced to Marianna Evans yet, so I didn't know the Morton Collection existed. I called some friends instead. One of them told me that my best chance of stumbling upon a large trove of centuries-old specimens was to fly to Paris and wait beside a set of trash cans along Rue Bonaparte. My guides would be waiting there on Tuesday night at seven p.m.

"This way," the leader said. The rusting steel door behind us moaned and squealed and the sliver of streetlight grew thinner until there was no more light at all, only fading echoes. One of the guides I was following lit up her high-powered headlamp, the two others cinched their backpacks, and went down the first stone steps of a spiral staircase that led into pure blackness.

The dead were downstairs. Six million of them, spread out in a labyrinth of halls, stalls, cathedrals, ossuaries, black rivers, and billionaire playrooms. There was the skull of Charles Perrault, the author of "Sleeping Beauty" and "Cinderella." A little deeper were the femurs of Antoine Lavoisier, the father of modern chemistry, and the ribs of Jean-Paul

Marat, the assassinated French Revolution leader and the subject of Jacques-Louis David's most morose painting. All these skulls, all these bones and millions of others, some dating back a thousand years, had been lying there, silently collecting dust underneath the Garden of Luxembourg in the heart of the Left Bank.

Leading the expedition was a woman in her early 30s with a purple-red mane that draped over a faded camouflage jacket. She was followed by another woman in a red pantsuit and a third in a fluorescent blue coat. They wore knee-high mud boots and overstuffed kitbags and looked like cast members from the all-female *Ghostbusters* reboot. I didn't know their real names and was told not to ask. These guides, I would learn, prefer to be anonymous.

At the foot of the staircase was a tunnel made of rough limestone walls. As we pushed deeper, the walls grew a little tighter, eventually forming a hexagonal shape—narrow at the feet, wide at the shoulders, and narrow again at the top. The tunnel had been built this way for efficiency, to allow ancient limestone miners to walk single file in as little space as possible. But the curious result was that the hallways were the shape of a coffin. Fitting, perhaps, because we'd just entered one of the largest graveyards on Earth.

For a thousand years, Parisians had buried their dead at the center of the city, mostly in a plot of land that became known as Holy Innocents' Cemetery. After hundreds of years of use, Holy Innocents' became overcrowded, and the dead were stacked in warehouses on top of one another. Those warehouses became overcrowded too, until the walls collapsed and spilled decomposing bodies into city streets. With nowhere to put the dead, Parisian authorities instructed limestone miners to dump them into wagons and wheel them into Paris's quarries. As new limestone quarries were dug out to build the Arc de Triomphe, the Louvre, and other great buildings, more bodies went underground. By the turn of the twentieth century, there were more than 170 miles of quarry tunnels filled with millions of skeletons.

The City of Paris offered a sanctioned tour of the quarries, called the Paris Catacombs, but that only covered a small portion. I'd come here to look inside the other 99 percent, where there were no tourists, descriptive plaques, ropes, lights, or rules. Where nothing was off-limits.

A group called "cataphiles" have been exploring the nether regions of this place since entering the quarries became illegal in 1955. They've found their way down through storm drains, manholes, and secret doorways along Rue Bonaparte. Some cataphiles had built private clubhouses within the limestone walls; others hosted weekly subterranean dance clubs. There was a rumor that a French billionaire chiseled out his own lavish apartment down there and hosted private parties, where guests do who-knows-what. Cataphiles made new discoveries all the time.

My guide, the woman with red-purple hair whom I'll call Red, had spent 15 years mapping these dirty tunnels. She was fascinated with the stories and history of the place. She told me earlier that she discovered a new ossuary an hour's hike from here in the crawl space of a cave. It was filled with a few thousand victims of a cholera epidemic that ravaged Paris in 1832. This was the time in Western history when small mouths, crooked teeth, and obstructed airways became the norm throughout much of industrial Europe. These were the skulls I was looking for.

We passed through halls, over puddles of stagnant water, and crawled face-to-butt like a human centipede through a kind of oversize rodent hole, until we reached a stack of wine bottles, cigarette pack wrappers, and dented beer cans. The walls were layered in decades of graffiti: the initials of two lovers, cartoon dicks, an obligatory 666. A few feet in front of us was a stack of what looked like kindling.

It wasn't kindling, not wood at all. It was a heap of femurs, humeri, sternums, ribs, and fibulas. Bones, all human. This was the path to the secret ossuary.

. . .

By around 1500, the farming that had begun in Southwest Asia and the Fertile Crescent ten thousand years earlier took over the world. The human population grew to a half billion, 100 times what it had been at the dawn of agriculture. Life, at least for city dwellers, was miserable: streams of human waste gushed down city streets. Air was tainted by coal smoke and nearby rivers and lakes ran with blood, fat, hair, and acids from manufacturing runoff. Infections, disease, and plague were a constant menace.

In these societies, for the first time in history, humans could spend their entire lives eating nothing but processed food—nothing fresh, nothing raw, nothing natural. Millions did. Over the next few centuries, food would become more and more refined. Advances in milling removed the germ and bran from rice, leaving only the starchy white seed. Roller mills (and, later, steam mills) ripped the germ and bran from wheat, leaving only a soft, white flour. Meats, fruits, and vegetables were canned and bottled. All these methods extended the shelf life of foods and made them more accessible to the public. But they also made foods mushy and soft. Sugar, which was once a prized commodity of the wealthy, became increasingly common and cheap.

This new, highly processed diet lacked fiber and the full spectrum of minerals, vitamins, amino acids, and other nutrients. As a result, urban populations would grow sicker and smaller. In the 1730s, before the onset of industrialization, the average Briton stood about five-seven. Within a century, population shrank two inches, to less than five-five.

The human face began rapidly deteriorating, too. Mouths shrank and facial bones grew stunted. Dental disease became rampant, and the incidence of crooked teeth and jaws increased tenfold in the Industrial Age. Our mouths got so bad, so overcrowded, that it became common to have teeth removed altogether.

The sideways grin of the Dickensian street urchin was not the afflic-
tion of only a few sad, impoverished orphans—the upper classes suffered, too. "The better the school, the worse the teeth," a Victorian
dentist observed. Breathing problems skyrocketed.

.   .   .

Back in the quarries, Red led me through the narrow opening of the
ossuary, across rocks and bones and broken bottles. She told me how
the cholera epidemic of the early 1800s killed close to 20,000 people.
The authorities had nowhere to put the dead, so they dug a big hole in
Montparnasse Cemetery and buried them with quicklime to disintegrate
the flesh. The ossuary was located at the bottom of that hole.

About ten more minutes of crawling and we reached it, a room surrounded by piles of bones and skulls. I had expected this place to have a
horror-show creepiness, but it never materialized. Instead, entering
there, surrounded by remnants of all these ancient lives, there was only
a long and heavy stillness, like the sound of a rock dropped in a well after
the echoes fade away.

Red and the cataphiles placed candles on the skulls and pulled cans
of beer and groceries from their packs. I turned and wormed myself
deeper into the chasm, pulling my body along the floor until it felt like my
chest might be stuck between two huge boulders. At one point, I considered that if any of us might suddenly get trapped in here, if we were to
break a leg, panic, or lose our way, there was a good chance we'd never
make it back out. Our skulls would join the millions of others lining
these walls, becoming candle holders for cataphiles in some future world.

Onward and inward, another wiggle and another yank, and I was in
the thick of it—hundreds of more skulls in all directions. These people
had been city dwellers, and they'd very likely relied on the same highly
processed industrial foods. To my eye, their skulls all appeared lopsided,

too short, their arches V-shaped, and stunted in some way. I stayed for a while bathing in them, inspecting them, feeling them, comparing them.

Admittedly, I was very much a novice at inspecting skeletons, and perhaps some of the jaws and other pieces were mismatched. Nonetheless, there was such a clear difference in the shape and symmetry of these specimens compared to the dozens of hunter-gatherers and other ancient indigenous populations I'd seen in books and websites before coming here. These were the Patient Zeros of the modern Industrial human mouth.

"*Voulez-vous manger quelque chose?*" Red said, her words echoing off the bare walls. I shimmied back beneath the crawl space and joined the group. They were smoking, sharing swigs of arak from a flask, and passing around snacks in the flickering glow of candlelight. Red pulled off a chunk of soft white bread and a slice of plastic-wrapped cheese and handed it to me. Beneath the stare of all those age-old eyeholes, I took a bite and mashed it in my crooked mouth a couple of times.

. . .

Researchers have suspected that industrialized food was shrinking our mouths and destroying our breathing for as long as we've been eating this way. In the 1800s, several scientists hypothesized that these problems were linked to deficiencies of vitamin D; without it, bones in the face, airways, and body couldn't develop. Others thought the lack of vitamin C was the culprit. In the 1930s, Weston Price, the founder of the National Dental Association research institute, decided it wasn't one specific vitamin or another, but all of them. Price set out to prove his theory. But unlike his predecessors, he wasn't interested in the causes of our shrinking mouths and deforming faces. He was interested in finding a cure.

"Since we have known for a long time that savages have excellent teeth and that civilized men have terrible teeth, it seems to me that we have been extraordinarily stupid in concentrating all of our attention upon the task of finding out why all our teeth are so poor, without ever bothering to learn why the savage's teeth are good," wrote Earnest Hooton, a Harvard anthropologist who supported Dr. Price's work.

Over a decade starting in the 1930s, Price compared the teeth, airways, and general health of populations around the world. He examined indigenous communities whose members were still eating traditional foods, comparing them to other members in the same community, sometimes the same family, who had adopted a modern industrialized diet. He traveled to a dozen countries, often in the company of his wife, and compiled more than 15,000 print photographs, 4,000 slides, thousands of dental records, saliva and food samples, films, and a library of detailed notes.

The same story played out no matter where he went. Societies that replaced their traditional diet with modern, processed foods suffered up to ten times more cavities, severely crooked teeth, obstructed airways, and overall poorer health. The modern diets were the same: white flour, white rice, jams, sweetened juices, canned vegetables, and processed meats. The traditional diets were all different.

In Alaska, Price found communities who ate seal meat, fish, lichen, and not much else. Deep inside Melanesian islands he found tribes whose meals consisted of pumpkins, pawpaws, coconut crabs, and sometimes long pigs (humans). He flew to Africa to study the nomadic Maasai, who subsisted mostly on cow's blood, some milk, a few plants, and a bite of steak. Then he traveled to Central Canada and studied indigenous tribes who suffered through winters when the temperature, according to Price's notes, could reach 70 degrees below zero (Fahrenheit) and whose only food was wild animals.

Some cultures ate nothing but meat, while others were mostly vegetarian. Some relied primarily on homemade cheese; others consumed no dairy at all. Their teeth were almost always perfect; their mouths were exceptionally wide, nasal apertures broad. They suffered few, if any, cavities and little dental disease. Respiratory diseases such as asthma or even tuberculosis, Price reported, were practically nonexistent.

While the foods in these diets varied, they all contained the same high amounts of vitamins and minerals: from one and a half to 50 times that of modern diets. All of them. Price became convinced that the cause of our shrinking mouths and obstructed airways was deficiencies not of just D or C but *all* essential vitamins. Vitamins and minerals, he discovered, work in symbiosis; one needs the others to be effective. This explained why supplements could be useless unless they're in the presence of other supplements. We needed all these nutrients to develop strong bones throughout the body, especially in the mouth and face.

In 1939, Price published *Nutrition and Physical Degeneration*, a 500-page doorstop of data collected during his travels. It was "a masterpiece of research," according to the *Canadian Medical Association Journal*. Earnest Hooton called it one of the "epochal pieces of research." But others hated it, and vehemently disagreed with Price's conclusions.

It wasn't Price's facts and figures, or even his dietary advice that made them bristle. Most of what he'd discovered about the modern diet had already been verified by nutritionists years earlier. But some complained that Price overreached, that his observations were too anecdotal and his sample sizes too small.

None of it mattered. By the 1940s, the idea of spending hours a day preparing meals of fish eyes and moose glands, raw roots and cow blood, coconut crabs and pig kidneys, seemed outdated and quaint. It was also way too much work. Many people moved to cities to get away from those foods and the grimy lifestyle that came with them.

Price, it turned out, was also only half right. Yes, vitamin deficiencies might explain why so many people eating industrialized foods were sick; they might explain why so many were getting cavities and why their bones were growing thin and weak. But they couldn't fully explain the sudden and extreme shrinking of the mouth and blocking of airways that swept through modern societies. Even if our ancestors consumed a full spectrum of vitamins and minerals every day, their mouths would still grow too small, teeth would come in crooked, and airways would become obstructed. What was true for our ancestors was also true for us. The problem had less to do with what we were eating than how we ate it.

Chewing.

It was the constant stress of chewing that was lacking from our diets—not vitamin A, B, C, or D. Ninety-five percent of the modern, processed diet was soft. Even what's considered healthy food today—smoothies, nut butters, oatmeal, avocados, whole wheat bread, vegetable soups. It's all soft.

Our ancient ancestors chewed for hours a day, every day. And because they chewed so much, their mouths, teeth, throats, and faces grew to be wide and strong and pronounced. Food in industrialized societies was so processed that it hardly required any chewing at all.

This is why so many of those skulls I'd examined in the Paris ossuary had narrow faces and crooked teeth. It's one of the reasons so many of us snore today, why our noses are stuffed, our airways clogged. Why we need sprays, pills, or surgical drilling just to get a breath of fresh air.

. . .

The cataphiles gathered their packs and bottles and cigarette butts from the ossuary, and I followed them back through the crawl spaces, across the fetid streams, up stone staircases, and out the secret door onto Rue Bonaparte. They scurried me past the police station and onto the Métro, where a trail of human bone dust followed me like breadcrumbs from Victor Hugo station back to a friend's apartment.

I left Paris mildly haunted. Not by the piles of bones in those underground warrens, but by the comprehensiveness of our folly. What looked like human progress—all that milling, mass distribution, and preservation of food—had horrible consequences.

Breathing slow, less, and exhaling deeply, I realized, none of it would really matter unless we were able to get those breaths through our noses, down our throats, and into the lungs. But our caved-in faces and too-small mouths had become obstacles to that clear path.

I spent a few days feeling sorry for humanity and then quickly set off in search of solutions. There had to be procedures, manipulations, or exercises that could reverse the past few centuries of damage from soft and mushy industrialized food. There had to be something that could help me with my own obstructed airways, and the wheezing, respiratory problems, and congestion I'd often experienced.

I started by visiting modern medical offices, meeting with specialists who looked at the top of the nose and worked down from there.

———————————

Dr. Nayak, the Stanford nose surgeon, told me during our earliest meeting that most of the nasal unblocking work he does involves turning "a one-lane highway into a two-lane highway." If a sink is plugged, we find a way to clear it safely and rapidly. Sometimes we'll use Drano for a minor clog; if that doesn't work, we'll call in a plumber. The nose tends to work in the same way. Sprays, rinses, and allergy medications can help quickly clear minor congestion, but for more serious chronic obstruction, we'll need a surgeon to plumb the path. I heard this analogy a lot.

Should I, or anyone else, develop a chronic mild nasal obstruction at any point in the future, Nayak first recommended a "Drano" approach in the form of a saline nasal rinse, sometimes with a low-dose steroid spray, a treatment that costs next to nothing and can be self-administered. He has also prescribed a topical rinse spiked with higher-dose steroids for patients on the path to reconstructive nasal surgery and found that 5 to 10 percent of patients no longer felt the need for further treatment.

Should obstruction become more stubborn sinus infections, Nayak might offer a patient a balloon. In this procedure, he inserts a small balloon into the sinuses and carefully inflates it. Balloon sinuplasty, as it's commonly called, creates more space for mucus and infection to pass out, and air and mucus to pass in. In one unpublished case-control study, Nayak found that, of the 28 selected sinusitis patients who received the procedure, 23 needed no other treatment.

Sometimes the nostrils are the problem, not the sinuses. Nostrils that are too small or that collapse too easily during an inhale can inhibit the free flow of air and contribute to breathing problems. This condition is so common that researchers have an official name for it, "nasal valve collapse," and an official measurement, called the Cottle's maneuver. It involves placing an index finger on the side of one or both nostrils and gently pulling each cheek outward, lightly spreading the nostrils open. If doing this improves the ease of nasal inhales, there's a chance that the nostrils are too small or thin. Many people with this condition receive minimally invasive surgery, or use adhesive strips called Breathe Right or nasal dilator cones.

If these simpler approaches fail, the drills come out. About three-quarters of modern humans have a deviated septum clearly visible to the naked eye, which means the bone and cartilage that separate the right and left airways of the nose are off center. Along with that, 50 percent of us have chronically inflamed turbinates; the erectile tissue lining our sinuses is too puffed up for us to breathe comfortably through our noses.

Both problems can lead to chronic breathing difficulties and an increased risk of infections. Surgery is highly effective in straightening or reducing these structures, but Nayak warned that it needs to be done carefully and conservatively. The nose, after all, is a wondrous, ornate organ whose structures work as a tightly controlled system.

The vast majority of nasal surgeries are successful, Nayak told me. Patients wake up, take the splints out and bandages off. No more congestion. No more sinus headaches. No more mouthbreathing. They are on their way to a new life, breathing better than they ever did before.

But not all of them. If surgeons drill out or remove too much tissue, especially the turbinates, the nose can't effectively filter, humidify, clean, or even sense inhaled air. For this small and unfortunate group of patients, each breath comes in too quickly, a hideous condition called empty nose syndrome.

I interviewed several empty nose sufferers, seeking to understand their condition. I talked for months with Peter, a laser technician who worked in the aeronautics industry in Seattle. He'd scheduled a surgery, hoping to clear up some minor obstruction, and, against his permission, had 75 percent of his turbinates removed over two procedures. Within days of the first, he felt a sense of suffocation. He couldn't sleep. The surgeons convinced Peter they hadn't removed enough, so they went back in. The second surgery made things much worse. For years later, each breath Peter took shot a bolt of pain to his brain, as if it had been delivered from an air pump. Doctors told Peter nothing was wrong; they prescribed antidepressants and suggested regular exercise. At one point he contemplated suicide.

I traveled thousands of miles to meet with other empty nose syndrome sufferers, trying to get context on this confounding and cruel condition. I learned that Peter's story really wasn't unique; there were thousands of people in countries all over the world afflicted with even worse fates. They too had been living successful and happy lives. They were students, corporate executives, or moms with chronic sinusitis who were tired of taking antibiotics. They'd had their noses examined and been ushered into surgery. They woke up to find much of their nasal structures removed, sometimes against their permission. With every breath they felt more breathless and anxious, a pain and dryness coating their airways. Many were reassured that any discomfort would get better in time. It often got worse.

Hundreds of people with empty nose syndrome told this same

overlapping story: they complained about anxiety, insomnia, and depression. Many were forced to quit lucrative jobs. Their doctors, families, and friends couldn't understand. Having access to more air, more quickly, could only be an advantage, they said. But we know now that the opposite is more often true.

Five percent of Nayak's patients in the past six years—nearly 200 people from 25 states and 7 countries—have come to Stanford to understand if and how empty nose syndrome is affecting them, and what procedures might help them breathe normally again. If they pass a rigorous screening test, Nayak will go into their noses and reconstruct the soft tissues and cartilage that had been taken out.

One estimate says that up to 20 percent of patients who got their lower (inferior) turbinates removed were at risk of eventually suffering some degree of empty nose syndrome, although Nayak believes these numbers are grossly overinflated. The number of patients complaining of breathing difficulties after more minor procedures is certainly far lower, but even if they represented 1 percent of 1 percent, the empty nose stories spooked me enough to explore other options before I ever went under the knife to fix my obstructed breathing.

So I dug a little deeper, a little lower, into the mouth.

. . .

Sleep apnea and snoring, asthma and ADHD, are all linked to obstruction in the mouth. There are no professionals who spend more time looking in the mouth than dentists. I talked to a half-dozen who specialize in procedures to remove obstacles. Here's what they told me to look for.

If you go to a mirror, open your mouth, and look at the back of the throat, you'll see a fleshy tassel that hangs bat-like from the soft tissues. That's the uvula. In mouths least susceptible to airway obstruction, the uvula will appear high and clearly visible from top to bottom. The deeper the uvula appears to hang in the throat, the higher the risk of airway obstruction. In mouths that are most susceptible, the uvula may not be visible at all. This measurement system is called the Friedman tongue position scale, and it's used to quickly estimate breathing ability.

Next is the tongue. If the tongue overlaps the molars, or has "scalloping" teeth indentations on its sides, it's too large and will be more apt to clog the throat when you lie down to sleep.

Farther down is the neck. Thicker necks cramp airways. Men with neck circumferences of more than 17 inches, and women with necks larger than 16 inches, have a significantly increased risk of airway obstruction. The more weight you gain, the higher your risk of suffering from snoring and sleep apnea, although body mass index is only one of many factors. Weight lifters frequently deal with sleep apnea and chronic breathing problems; instead of layers of fat, they have muscles crowding the airways. Plenty of rail-thin distance runners and even infants suffer, too.

That's because the blockage doesn't start with the neck, uvula, or tongue. It starts with the mouth, and mouth size is indiscriminate. Ninety percent of the obstruction in the airway occurs around the tongue, soft palate, and tissues around the mouth. The smaller the mouth is, the more the tongue, uvula, and other tissues can obstruct airflow.

There are various ways to improve matters for airway obstructions. Dr. Michael Gelb is a renowned New York dentist who specializes in treating snoring, sleep apnea, anxiety, and other breathing-related problems. "I see her, this same patient, every day," he told me when I visited his Madison Avenue clinic in New York. Many of Gelb's patients, he said, don't fit the traditional mold. They are mid-30s, fit, successful. No health issues growing up, but in the last couple of years they'd experienced fatigue, bowel issues, headaches. Their ears hurt when they bite down. Their primary care doctors misdiagnose them and prescribe antidepressants, but the drugs don't work. So they try a continuous positive airway pressure mask, or CPAP, which forces bursts of air past the obstructed airways into the lungs.

CPAPs are a lifesaver for those suffering from moderate to severe sleep apnea, and the devices have helped millions of people finally get a good night's rest. But Gelb told me his patients have a hard time wearing them. Further, many don't have medically diagnosed sleep apnea; the data from sleep studies shows they're breathing during sleep just fine. Yet these people keep getting more tired, forgetful, and sick. These people may not register a sleep apnea problem, Gelb told me, but they all had a serious breathing problem. "By the time I get them, I'm dealing with the walking dead," he said.

Gelb and his colleagues sometimes remove tonsils and adenoids. This can be especially effective for children: 50 percent of kids with ADHD were shown to no longer have symptoms after having their adenoids and tonsils removed. But these effects can also be fleeting. Years after having tonsils removed, children can develop obstructions in the airways and all the problems that come with it. This is because neither adenoid/tonsil removal nor CPAP nor other procedures provide a satisfying long-term solution, because none deals with the core issue: a mouth that is too small for the face.

Gelb also offers treatments to correct head and neck posture, using various gizmos to force the jaw away from the airways. Most work. He showed me a gallery of patients who looked practically reborn after treatment. But I wasn't the walking dead—not yet, at least. The obstruction in my airways was much milder.

For me and the majority of the population, the best medicine, Gelb said, is preventative. It involves reversing the entropy in our airways so that we can avoid sleep apnea, anxiety, and all the chronic respiratory problems as we grow older. It involves expanding the too-small mouth.

. . .

The earliest orthodontics devices weren't intended to straighten teeth, but to widen the mouth and open airways. In the mid-1800s, a host of children were born with cleft palates and narrow V-shaped arches. Their mouths were so small that they had trouble eating, speaking, and breathing. Norman Kingsley, a dentist and sculptor, wanted to help them, so in 1859 he built a device that forced the jaw forward, creating room at the back of the mouth to open the throat. It worked well enough. By the 1900s, a French stomatologist named Pierre Robin was designing his own contraption.

Robin called it the "monobloc," and it consisted of a rubber retainer with a dowel screw that forced the upper palate to grow outward. Within just a few weeks, his patients' mouths grew larger and their breathing was significantly improved.

The monobloc kicked off a wave of other mouth-expanding devices that would be used for another benefit: straightening crooked teeth. Teeth will grow in naturally straight if they have enough room. Expanding devices returned the mouth to the width it was intended to be, offering a larger "playing field" for teeth. Expansion would remain a standard practice for the next 20 years, and would continue to be used throughout Europe for decades after that.

But the process of expanding a mouth took expertise and maintenance; results varied depending on the skill of the dentist. It didn't help that these devices were miserable and awkward to wear. For those patients with overbites, the most common problem in the mouth, few dentists could figure out how to move the bottom jaw forward, so they instead began working on ways to move the top of the mouth back.

By the 1940s, it became standard practice for dentists to extract teeth then crane back the remaining top teeth with headgear, braces, and other orthodontic devices. Fewer teeth were easier to handle and offered more consistent results. By the 1950s, tooth extractions—two, four, even six at a time—and retractive orthodontics were routine in the United States.

There was a glaring problem with this approach: removing teeth and pushing remaining teeth backward only made a too-small mouth smaller. A smaller mouth might be easy for dentists to manage, but it also offered less room to breathe.

A few months, or years, after their mouths were compressed with braces and headgear, some patients would complain about breathing difficulties like snoring, sleep apnea, hay fever, and asthma that they'd never had before. When they bit down, they noticed a clicking sound at the back of their jaws, along the temporomandibular joint. Some began to look different, their faces growing longer, flatter, and less defined.

These patients might have represented only a small percentage. But enough showed the same breathing problems, chewing problems, and downward facial growth that, in the late 1950s, a former biplane pilot, semiprofessional Formula 1 driver, and British facial surgeon and dentist named Dr. John Mew took notice.

Mew began measuring the faces and mouths of young patients who'd gotten extractions, and compared them with patients who'd gotten expansion treatment. Brothers and sisters measured against their siblings, even sets of identical twins. Over and over again, the children who'd had teeth removed and had undergone retractive orthodontics suffered from the same stunted mouth and facial growth. As they grew up, and the rest of their bodies and heads grew larger, their mouths were forced to stay the same size. This mismatch created a problem at the center of the face: eyes would droop, cheeks would puff up, and chins would recess. The more teeth these patients had extracted, the longer they wore braces and other devices, the more obstruction seemed to develop in their airways.

Mew called the pattern "sadly a common sequel to fixed orthodontic treatment."

In a strange twist, he found that the devices invented to fix crooked teeth caused by too-small mouths were making mouths smaller and breathing worse.

Mew wasn't alone. Several other dentists had come to the same conclusion, publishing scientific papers on the subject. Mew did his own studies, taking hundreds of measurements and before-and-after photos of his patients. He even conducted biochemical analysis of the cell structure of the lips. All of which, he claimed, clearly proved how the combination of extractions and retractive orthodontics hindered forward facial growth and breathing. He served as president of the Southern Counties Branch of the British Dental Association and used his influence to petition administrators to run a full investigation.

Nobody did anything; nobody really cared. Instead, Mew would become one of the most divisive men in British dentistry, ridiculed as a "quack," "scammer," and "snake oil salesman." He was sued repeatedly to stop from practicing expansion and would eventually lose his license. As Mew approached the tenth decade of his life, it appeared he was going to follow the same trajectory as Stough, Price, and so many other pulmonauts: to die in obscurity, buried along with his research.

But something curious has happened in the last few years. Hundreds of leading orthodontists and dentists have come out in support of Mew's position, saying that, yes, traditional orthodontics were making breathing worse in half their patients. The strongest endorsement came in April 2018, when Stanford University Press published a 216-page monograph by famed evolutionary biologist Paul R. Ehrlich and Dr. Sandra Kahn, an orthodontist, detailing hundreds of scientific references that supported Mew's research. In a short time, Mew's outlier theories started entering the mainstream.

"In ten years, nobody will be using traditional orthodontics," Gelb told me. "We'll look back at what we've done and be horrified." This is what Mew had been saying for the past half century. The rebellion within orthodontics eventually led to the formation of a professional organization called the Academy of Orofacial Myofunctional Therapy.

This group, I'd learned, is much more interested in fixing the problem of undersize mouths than blaming those who contributed to it. There are too many variables, and too many guilty parties, they argued. As with so many fixes I'd come across, Mew and the others discovered that the tools they needed to remove airway obstruction, to restore the function in that too-small mouth, were created long ago by observant scientists whose research was accepted as the standard—and then, for one reason or another, forgotten.

. . .

I visited John Mew two weeks after my expedition into the Paris quarries. I arrived at an empty train station in East Sussex and an hour later was in the passenger seat of a Renault minivan. Mew was at the wheel, driving twice the speed limit down a country road canopied in trees in the posh suburb of Broad Oak, about 90 minutes east of London.

"I've come under incredible resistance the whole way," he told me, scraping the passenger-side door against an overgrown thicket as we whizzed down a one-way street. "But the science is clear, the facts are clear, the evidence is everywhere. There's really no way they can keep stopping it."

It was a Sunday afternoon, and Mew's only plans were to meet with me and have his children over for tea, but he'd dressed in a three-piece houndstooth suit with a white shirt and a rep tie from his preparatory school, which he attended 75 years ago. We veered onto a gravel driveway and over a small bridge, then parked in the shadow of a stone turret.

I had heard that Mew lived in a "castle," and was expecting something castle-esque, with painted concrete and vinyl siding. But every detail of this place appeared strikingly real, from the moss-covered roof to the blackwater moat. Mew killed the motor, grabbed his cane, and led me through dark hallways to a kitchen of black wood cabinets and copper pots.

For several hours we sat beside a roaring hearth, where I heard about how Mew built this castle, doing much of the work himself over the course of a decade when he was in his late 70s. I also heard about Mew's various devices to expand mouths.

His most renowned invention was the Biobloc, a modified version of Pierre Robin's monobloc. Mew used it on hundreds of his own patients; hundreds of orthodontists still use it today. A 2006 peer-reviewed study of 50 children showed that the Biobloc expanded airways by up to 30 percent over the course of six months.

I came here because I'd become interested in expanding my own too-small mouth and opening my too-small airways. But Mew told me his device works best for children age 5 to 9, whose bones and faces were still developing and easily moldable. For me, that was several lifetimes ago.

Mew's son, Mike, who is also a dentist, joined the conversation. Mike was tanned, tall, and lanky, with piercing brown eyes, dressed in fashion jeans and a tight-fitting sweater. He explained that the first step to improving airway obstruction wasn't orthodontics but instead involved maintaining correct "oral posture." Anyone could do this, and it was free.

It just meant holding the lips together, with teeth almost touching (molars two to three millimeters apart) and the tongue on the roof of the mouth. Hold the head up perpendicular to the body and don't kink the neck. When sitting or standing, the spine should form a slight J shape—straight until it reaches the small of the back, where it naturally curves outward. While maintaining this posture, we should always breathe slowly through the nose into the abdomen.

Our bodies and airways are designed to work best in this posture, both Mews agreed. Look at any Greek statue, or a drawing by Leonardo, or an ancient portrait. Everyone shared this J shape. But if we look around public spaces today, it's obvious that most people have shoulders hunched forward, neck extended outward, and an S-shaped spine. "A bunch of village idiots, that's what we've become," shouted Mike. He then assumed this "idiot" position, inhaled a few short, puffy, open-mouth breaths, and looked around dumbly. "It's bloody killing us!"

Many of us adopted this S posture not because of laziness but because our tongues don't fit properly in our too-small mouths. Having nowhere else to go, the tongue falls back into the throat, creating a mild suffocation. At night, we choke and cough, attempting to push air in and out of this obstructed airway. This, of course, is sleep apnea, and a quarter of Americans suffer from it.

By day, we unconsciously attempt to open our obstructed airways by sloping our shoulders, craning our necks forward, and tilting our heads up. "Think of someone who is unconscious and about to receive CPR," Mike said. The first thing a medic does is tilt the head back to open the throat. We've adopted this CPR posture all the time.

Our bodies hate this position. The weight of the sloping head stresses the back muscles, leading to back pain; the kink in our necks adds pressure to the brain stem, triggering headaches and other neurological problems; the tilted angle of our faces stretches the skin down from the eyes, thins the upper lip, pulls flesh down on the nasal bone. Because "the village idiot stare" doesn't sound scientific, Mike calls this posture "cranial dystrophy." He claims it affects about 50 percent of the modern population, including Mark Zuckerberg, the founder of Facebook.

In January 2018, Mike uploaded a YouTube video warning Zuckerberg that he was going to die ten years early if he didn't correct his cranial dystrophy posture. The message was viewed more than 9,000 times before it was deleted.

Along with maintaining the correct oral posture, Mike recommended a series of tongue-thrusting exercises, which he says can train us out of the "death pose" and make breathing easier. The tongue is a powerful muscle. If its force is directed at the teeth, it can throw them out of alignment; if it's directed at the roof of the mouth, Mike believed it might help expand the upper palate of the mouth and open up the airways.

The exercise, which Mike's hordes of social media fans call "mewing," has been popularly adopted as "a new health craze." After a few months, mewers have claimed their mouths expanded, jaws became more defined, sleep apnea symptoms lessened, and breathing became easier. Mike's own instructional video on mewing has been viewed a million times.

It's difficult to convey mewing without seeing it, but the gist is to push the back of the tongue against the back roof of the mouth and move the rest of the tongue forward, like a wave, until the tip hits just behind the front teeth. I tried it a few times. It felt awkward, like I was holding back vomit. Mike demonstrated it for me. It looked like he was holding back vomit.

It was then—mewing in unison with another grown man in a home-made castle, clumps of human bone dust still caked in the eyelets of my boots—that I realized the quest to discover the lost art of breathing was going to be a bit of a shit show.

But I kept at it, mewing all the way back out through the arched hallways and into a moonless night, thinking how much more I'd enjoy the practice if I understood why it worked.

. . .

Which is how I found myself at a final stop, on a dental examination chair a few blocks south of Grand Central Terminal. Dr. Theodore Belfor was hunched over me, wearing a short-sleeve shirt, gray slacks, and wingtips, his shaved head gleaming under the examination lights. He was cleaning a dental impression mold in the sink and explaining how human evolution is no longer based on the survival of the fittest, an echo of what I'd hear from Marianna Evans. He was also describing how my mouth was a total mess because of it.

Belfor was another dentist with big ideas about how humans lost the ability to breathe. And like the Mews and Gelb, he had big ideas on how to fix it.

"Hold still," he said in a thick Bronx accent as he reached his big hands into my mouth. "Narrow arch, crowding, recessed mandible—you've got it all. Very typical."

In the 1960s, after graduating from New York University College of Dentistry, Belfor was sent to Vietnam to work as the sole dentist and mouth surgeon for 4,000 soldiers in the 196th Light Infantry. He had no oversight and was able to improvise, invent, and devise novel solutions to what were often disastrous problems. "I really learned how to put faces back together," he said, chuckling.

He returned to New York and was offered a job working on perform-ing artists. These singers, actors, and models needed straight teeth but couldn't be seen with braces. A colleague introduced him to an old monobloc-like device. After a few months of using it, opera singers began hitting higher notes and chronic snorers slept peacefully for the first time in years. Everyone had straighter teeth and reported breathing better. Some in their 50s and 60s noticed the bones in their mouths and faces growing wider and more pronounced the longer they wore the devices.

The results stunned Belfor. He'd been taught, like everyone else, that bone mass (just as with lung size) only decreases after the age of 30. Women will suffer much more bone loss than men, especially after menopause. By the time a woman reaches 60, she'll have lost more than a third of her bone mass. If she lives to 80, she'll have as much bone as she had when she was 15. Eating well and getting exercise can help to stave off the deterioration, but nothing can stop it.

It's most apparent in our faces. Sagging skin, baggy and hollow eyes, and sallow cheeks all result from bone disappearing and flesh having nowhere to go but down. As bone degrades deeper in the skull, soft tissues at the back of the throat have less to hang on to, so they can droop too, which can lead to airway obstruction. This bone loss partly explains why snoring and sleep apnea often grow worse the older we get.

After decades of experimentation and collecting case studies, of seeing his patients' mouths and faces grow younger the older they got, Belfor decided that the conventional science of bone loss was, in his words, "total bullshit."

"Clamp your teeth," he told me. I did and felt stress in my jaw that extended all the way back to my skull. What I was feeling was the power of the masseter, the chewing muscle located below and in front of the ears. It's the strongest muscle in the body relative to its weight, exerting up to 200 pounds of pressure on the back teeth.

Belfor then had me run my hands along my skull until I felt the web of cracks and ridges, called sutures. Sutures spread apart throughout our lives. This spreading allows the skull bone to flex and expand to double its size from infancy to adulthood. Inside these sutures, the body creates stem cells, amorphous blanks that shift form and become tissues and bones depending on what our bodies need. Stem cells, which are used throughout the body, are also the mortar that binds the sutures together and that grows new bone in the mouth and face.

Unlike other bones in the body, the bone that makes up the center of the face, called the maxilla, is made of a membrane bone that's highly plastic. The maxilla can remodel and grow more dense into our 70s, and likely longer. "You, me, whoever—we can grow bone at any age," Belfor told me. All we need are stem cells. And the way we produce and signal stem cells to build more maxilla bone in the face is by engaging the masseter—by clamping down on the back molars over and over.

Chewing. The more we gnaw, the more stem cells release, the more bone density and growth we'll trigger, the younger we'll look and the better we'll breathe.

It starts at infancy. The chewing and sucking stress required for breastfeeding exercises the masseter and other facial muscles and stimulates more stem cell growth, stronger bones, and more pronounced airways. Until a few hundred years ago, mothers would breastfeed infants up to two to four years of age, and sometimes to adolescence. The more time infants spent chewing and sucking, the more developed their faces and airways would become, and the better they'd breathe later in life. Dozens of studies in the past two decades have supported this claim. They've shown lower incidence of crooked teeth and snoring and sleep apnea in infants who were breastfed longer over those who were bottle-fed.

"Now scoot down and put your head back," Belfor said, pointing the dental impression tray toward my open mouth. The mold he was about to take would be used to fit me with a Homeoblock, an expanding device Belfor invented in the 1990s. It's a pink acrylic thing wrapped in gleaming metal wires that looks no different from any other retainer. Except the Homeoblock wasn't designed to straighten teeth. Like the first functional orthodontic devices created by Norman Kingsley and Pierre Robin, its purpose is to expand the mouth and make breathing easier. Along the way, it stimulates the stress of chewing whenever the wearer chomps down, so they won't have to spend three to four hours gnawing on bones and bark like our ancient relatives.

Belfor's patients—who included Richard Gere's body double, a middle-aged housewife from Phoenix, a 79-year-old New York socialite, and hundreds of others—shared profound results. Belfor showed me their before-and-after CAT scans when I first got to his office. They had obstructed throats in the shots before; more-open airways and loads of new bone six months later. It was as if these patients were the dental equivalent to Dorian Gray.

"Now open your mouth wider and say *aaahhhhhhh*," Belfor said.

. . .

The chew-airway connection, like so much else breath-related, was old news. As I dug through a century of scientific papers on the subject over several months, I felt like I was trapped in a respiratory research spin cycle. Different scientists, different decades; the same conclusions, the same collective amnesia.

James Sim Wallace, a renowned Scottish doctor and dentist, published several books about the deleterious effects of soft foods on our mouths and breathing. "An early soft diet prevents the development of the muscle fibers of the tongue," he wrote more than a century ago, "resulting in a weaker tongue which [cannot] drive the primary dentition out into a spaced relationship with fully developed arches which will lead to more crowding of the permanent teeth."

Wallace's contemporaries began taking measurements of patients' mouths and comparing them to skulls that dated to before the Industrial Revolution. The palates of the ancient skulls measured an average of 2.37 inches. By the late nineteenth century, mouths had shrunk to 2.16 inches. No one was disputing these observations. "That the human jaw is gradually becoming smaller is a fact which is universally recognized," Wallace noted. That didn't stop this research from being ignored for the next hundred years.

By 1974, though, a shaggy-haired 26-year-old anthropologist at the Smithsonian National Museum of Natural History picked up the baton. His name was Robert Corruccini, and he'd write or contribute to 250 research papers and a dozen books on the topic. Corruccini traveled the world and examined thousands of mouths and diets, from Pima Native Americans to urban populations of Chinese immigrants, rural Kentuckians to Australian Aboriginals. He even conducted animal studies, feeding a group of pigs a diet of hard-pelleted chow and others the identical chow softened with water. The same food, the same vitamins; only the texture had changed.

People, pigs, whatever. Whenever they switched from harder foods to soft foods, faces would narrow, teeth would crowd, jaws would fall out of alignment. Breathing problems would often follow.

Fifty percent of the modern human population would show this "malocclusion" within the first generation of switching to soft and processed foods; by the second generation, 70 percent; by the third, 85 percent.

By the fourth, well, look around. That's us, now. Some 90 percent of us have some form of malocclusion.

Corruccini presented his groundbreaking data to dental conferences around the United States, calling crooked teeth a "disease of civilization." There was a lot of interest at first. "A really polite reception," he said. "But nothing really changed."

Today, the official website of the U.S. National Institutes of Health attributes the causes of crooked teeth and other deformations of the airway "most often to heredity." Other causes include thumb-sucking, injury, or "tumors of the mouth and jaw."

There is no mention of chewing; no mention of food at all.

. . .

Belfor collected his own library of data over two decades. He had case studies and charts and graphs showing how his patients were regrowing bones and opening their airways. But he too was universally ignored, and often ridiculed. After one lecture at his alma mater, several colleagues claimed he had faked his data and Photoshopped his X-rays. "You cannot grow bone past 30," they reprimanded him, time and again.

Belfor and Corruccini are still waiting for their Mew moment, when the establishment starts to come around. In the meantime, I've come around.

Exactly a year to the week after I began wearing Belfor's retainer, I visited a private radiology clinic in downtown San Francisco and had my airways, sinuses, and mouth rescanned. Belfor sent the results to Analyze-Direct at the Mayo Clinic to study what had happened to my face and airways.

The results were stunning. I had gained 1,658 cubic millimeters of new bone in my cheeks and right eye socket, the equivalent volume of five pennies. I'd also added 118 cubic millimeters of bone along my nose, and 178 along my upper jaw. My jaw position became better aligned and balanced. My airways widened and became firmer. The deposit of pus and granulation that had accumulated in my maxillary sinuses, likely the result of mild chronic obstruction, was completely gone.

Sure, it took weeks to get used to having a chunk of plastic in my mouth at night. Spit built up, my throat constricted, and my teeth ached. But like most discomforts in life, it got easier and less annoying the longer I did it.

As I write this, because of chewing and some widening of my palate, I am breathing more easily and freely than I ever remember. Other than that week and a half in which I purposely obstructed my nose in the Stanford experiment, I have suffered only one stuffy nose this year, when I came down with a cold. Even with my messed-up, middle-aged mouth and face, I'd managed to make real progress.*

"Nature seeks homeostasis and balance," Belfor told me on the phone

---

*Nobody needs a Homeoblock or retainer to get the bone-building and airway-expanding benefits of chewing. Hard, natural foods and chewing gum likely work just as effectively. Marianna Evans recommends her patients chew gum for a couple of hours a day. I too followed this advice and some days I would chomp on an extremely hard type of Turkish gum called Falim, which came in flavors like carbonate and mint grass. The stuff tasted pretty crude, but it offered a workout and delivered results.

in one of our dozens of conversations since we first met. "You were out of balance. Just look at the scans. Nature corrected you by adding a tremendous amount of bone to your face—the proof is in the pudding."

This is what I learned at the end of this long and very strange trip through the causes and cures of airway obstruction. That our noses and mouths are not predetermined at birth, childhood, or even in adulthood. We can reverse the clock on much of the damage that's been done in the past few hundred years by force of will, with nothing more than proper posture, hard chewing, and perhaps some mewing.

And with the obstruction out of the way, we can finally get back to breathing.

Part Three

○——○

BREATHING+

Eight

MORE, ON OCCASION

o——o

The morning after our celebratory "last" supper, Olsson and I hop back in my car and drive down to Stanford for our final inspection with Dr. Nayak. We are rescanned, reprodded, repoked, and repeppered with questions. The same tests we ran through ten days ago, and ten days before that. The data for both stages of the experiment, we're told, will be available later this month. We are free to breathe, free to go.

For Olsson, that means heading back to Sweden. For me, it means further exploration to the outer limits of breath.

.   .   .

The techniques I'll pursue from this point forward will not hold to the slow-and-steady style. They are *not* accessible to everyone, everywhere. You can't practice them while flipping through the pages of this book.

Several take a long time to master, require concerted effort, and can be uncomfortable.

Pulmonary medicine has many scary names for what these more extreme techniques can do to the body and mind: respiratory acidosis, alkalosis, hypocapnia, sympathetic nervous system overload, extreme apnea. Under normal circumstances, these conditions are considered damaging and would require medical care.

But something else happens when we practice these techniques *willingly*, when we consciously push our bodies into these states for a few minutes, or hours, a day. In some cases, they can radically transform lives.

Collectively, I'm calling these potent techniques Breathing+, because they build on the foundation of practices I described earlier in this book, and because many require extra focus and offer extra rewards. Some involve breathing really fast for a very long time; others require breathing very slow for even longer. A few entail not breathing at all for a few minutes. These methods, too, date back thousands of years, vanished, then were rediscovered again at a different time in a different culture, renamed and redeployed.

At best, Breathing+ can offer a deeper view into the secrets of our most basic biological function. At worst, breathing this way can provoke heavy sweats, nausea, and exhaustion. This, I would learn, is all part of the process. It's the respiratory gauntlet required to get to the other side.

· · ·

As unlikely as it sounds, Civil War battlefields are where the first Breathing+ technique begins.

It was 1862 and Jacob Mendez Da Costa had just arrived at Turner's Lane Hospital in Philadelphia. The Union Army had suffered a humiliating defeat at Fredericksburg, Virginia, where twelve hundred men had been killed and more than 9,000 wounded. Soldiers were laid out in the hallways, bruised and bleeding on rows of cots, missing ears, fingers, arms, and legs.

Even those who hadn't seen military action were falling apart. They came into the hospital in droves, complaining of anxiety and paranoia, headaches, diarrhea, dizziness, and shooting pain in their chests. They sighed a lot. When the men tried to breathe, they'd huff and huff, but would never feel like they could catch a breath. These men showed no signs of physical damage; they had spent weeks or months preparing for battle but never saw any action. Nothing had *happened* to them. And yet each was incapacitated, hobbling beneath the whitewashed walls of the hospital, past the rows of screaming and suffering amputees, trying to find their way into Da Costa's care.

Da Costa was a glum-looking man with a bald head, lambchop sideburns, and tired, Portuguese eyes. He'd been born on the island of St. Thomas and spent years studying medicine in Europe with leading surgeons. He'd become a renowned expert in the maladies of the heart and had treated scores of men with myriad ailments. But he'd never seen anything like the soldiers at Turner's Lane.

He started the examinations by lifting the men's shirts and placing a stethoscope to their chests. The soldiers' heartbeats were manic, thumping up to 200 beats per minute, even though they were sitting still. Some breathed 30 or more times a minute, double the normal pace.

One typical patient was William C., a 21-year-old farmer who, after deployment, developed vicious diarrhea and a bluish tint to his hands. He complained of breathlessness. Henry H. had identical symptoms and shared William C.'s skinny build, with a narrow chest and a stooped spine. He too had enlisted in good health, then, without explanation, was immobilized. "The man did not look sick," wrote Da Costa. But his heart rate was "of irregular rhythm, some beats following each other in rapid succession."

Hundreds of men would come to see Da Costa over the next few years, with the same cluster of complaints, the same backstory. Da Costa would call the malady Irritable Heart Syndrome.

The syndrome was puzzling in another way: the symptoms came on, and then they would disappear. A few days, weeks, or months of rest and relaxation, and heartbeats would soften, digestive problems would abate. The men became normal again and they'd breathe normally again, too. Most would be sent back to war. The few still suffering would be placed in the "invalid corps" or shipped home to deal with the syndrome for the rest of their lives.

Da Costa recorded reams of data on these men, and he released a formal clinical study in 1871, which would become a landmark in the history of cardiovascular disease.

But Irritable Heart Syndrome wasn't confined to just the Civil War. The same symptoms would show up half a century later in 20 percent of soldiers who fought in World War I, a million soldiers in World War II, and hundreds of thousands more in Vietnam and the Iraq and Afghanistan wars. Doctors dreamed up new names for these problems along the way, believing they had discovered a new kind of illness. They told soldiers they suffered from shell shock, soldier's heart, post-Vietnam syndrome, and post-traumatic stress disorder. They considered the ailments to be psychological, some disturbance in the brain brought on by fighting. Soldiers often blamed exposure to chemicals or vaccines, although nobody really knew for sure.

Da Costa had his own theories. At Turner's Hospital, he suspected he was dealing with, in his words, "a disorder of the sympathetic nervous system."

It is the same disorder I'm feeling right now.

. . .

It's late morning and I'm splayed out on a yoga mat on the parched lawn of a roadside public park in the foothills of the Sierra Nevada mountains. There's a picnic table full of emergency medical technicians eating lunch to my right, an old man brown-bagging a tallboy beer on a bench to my left. Above me, the autumn sun is so clear and bright that it's blinding, even through squinted eyes. I take a huge, heaving breath into the pit of my abdomen and let it out. I've been doing this for the past few minutes, and I can feel beads of sweat erupting on my forehead and face. I've got another half hour to go.

"Twenty more!" yells the man standing over me. I can barely hear him through the thunder of big rigs shifting gears on the highway behind us. His name is Chuck McGee III, and he's a big dude with a sandy bowl cut, rainbow-lensed blade glasses, and cargo shorts that dangle just a few inches above white socks and dirt-baked sneakers. I've hired him for the day to help me redline my sympathetic nervous system with overbreathing.

So far it's working. My heart is beating violently. It feels like there's a rodent loose in my chest. I feel anxious and paranoid, sweaty and claustrophobic.

This must be the sympathetic overload. This must be Irritable Heart Syndrome coming on.

Breathing, as it happens, is more than just a biochemical or physical act; it's more than just moving the diaphragm downward and sucking in air to feed hungry cells and remove wastes. The tens of billions of molecules we bring into our bodies with every breath also serve a more subtle, but equally important role. They influence nearly every internal organ, telling them when to turn on and off. They affect heart rate, digestion, moods, attitudes; when we feel aroused, and when we feel nauseated. Breathing is a power switch to a vast network called the autonomic nervous system.

There are two sections of this system, and they serve opposite functions. Each is essential to our well-being.

The first, called the parasympathetic nervous system, stimulates relaxation and restoration. The mellow buzz you get during a long massage or the sleepiness you feel after a big meal happens because the parasympathetic nervous system sends signals to your stomach to digest and to the brain to pump feel-good hormones such as serotonin and oxytocin into your bloodstream. Parasympathetic stimulation opens the floodgates in our eyes and makes tears flow at weddings. It prompts salivation before meals, loosens the bowels to eliminate waste, and stimulates the genitals before sex. For these reasons, it's sometimes called the "feed and breed" system.

The lungs are covered with nerves that extend to both sides of the autonomic nervous system, and many of the nerves connecting to the parasympathetic system are located in the lower lobes, which is one reason long and slow breaths are so relaxing. As molecules of breath descend deeper, they switch on parasympathetic nerves, which send more messages for the organs to rest and digest. As air ascends through the lungs during exhalation, the molecules stimulate an even more powerful parasympathetic response. The deeper and more softly we breathe in, and the longer we exhale, the more slowly the heart beats and the calmer we become. People have evolved to spend the majority of waking hours—and all of our sleeping hours—in this state of recovery and relaxation. Chilling out helped make us human.

The second half of the autonomic nervous system, the sympathetic, has an opposite role. It sends stimulating signals to our organs, telling them to get ready for action. A profusion of the nerves to this system are spread out at the top of the lungs. When we take short, hasty breaths, the molecules of air switch on the sympathetic nerves. These work like 911 calls. The more messages the system gets, the bigger the emergency.

That negative energy you feel when someone cuts you off in traffic or wrongs you at work is the sympathetic system ramping up. In these states, the body redirects blood flow from less-vital organs like the stomach and bladder and sends it to the muscles and brain. Heart rate increases, adrenaline kicks in, blood vessels constrict, pupils dilate, palms sweat, the mind sharpens. Sympathetic states help ease pain and keep blood from draining out if we get injured. They make us meaner and leaner, so we can fight harder or run faster when confronted with danger.

But our bodies are built to stay in a state of heightened sympathetic alert only for short bursts, and only on occasion. Although sympathetic stress takes just a second to activate, turning it off and returning to a state of relaxation and restoration can take an hour or more. It's what makes food difficult to digest after an accident, and why men have trouble getting erections and women often can't experience orgasms when they're angry.*

For all these reasons, it seems odd and counterintuitive to willingly place yourself in an extended state of extreme sympathetic stress, and to do this every day. Why make yourself light-headed, anxious, and flaccid? And yet, for centuries, the ancients developed and practiced breathing techniques that did just this.

. . .

The stress-inducing breathing method that brought me to this roadside public park is called Inner Fire Meditation, and it's been practiced by Tibetan Buddhists and their students for the past thousand years. Its history begins around the tenth century AD, when a 28-year-old Indian man named Naropa got bored with domestic life. He divorced his wife,

---

*Sexual arousal is controlled by the parasympathetic system and is usually accompanied, or can be induced, by soft and easy breaths. Meanwhile, orgasms are a sympathetic response, and are often preceded by fast, short, and sharp breathing. We're attracted to eyes of mates with large pupils partly because pupils dilate—a sympathetic response—during orgasm.

packed a bag, and walked northeast until he was surrounded by stone towers, pavilions, temples, and blue lotus. This dazzling place was Buddhist University Nalanda, and thousands of scholars from throughout the East gathered there to study astronomy, astrology, and holistic medicine. A few sought enlightenment.

Naropa excelled in his courses, mastering the lessons of Sutra and the secret techniques of Tantra, which had been handed down from one master to another over the millennia. He set out into the Himalayas to put everything he'd learned into practice, living inside a cave on the banks of the Bagmati River in what is now Kathmandu, Nepal. The cave was cold. Naropa harnessed the power of his breath to keep himself from freezing to death. The practice became known as Tummo, the Tibetan word for "inner fire."

Tummo was dangerous. If used incorrectly, it could elicit intense surges of energy, which could cause serious mental harm. For that reason, it was reserved only for advanced monks and stayed in the Himalayas, locked up in the Tibetan monasteries for the next thousand years.

Fast-forward to the early 1900s, when a Belgian-French anarchist and former opera singer was making her way up to Tibet with soot on her face, yak fur woven into her hair, and a red belt around her head. Her name was Alexandra David-Néel, and she was in her mid-40s traveling alone through India—unheard-of at the time for a Western woman.

David-Néel had spent most of her life exploring different philosophies and religions. As a teenager, she'd hung out with mystics, starved herself, beaten herself, and followed diets used by ascetic saints. She was into Freemasonry and feminism and free love. But it was Buddhism that really fascinated her. She taught herself Sanskrit, then set out on a spiritual pilgrimage through India and Tibet that lasted 14 years. Along the way she happened into a cave high in the Himalayas, just as Naropa had done. It was there that a Tibetan holy man passed down the instructions for the superheating power of Tummo.

"[Tummo was] but a way devised by the Thibetan hermits of enabling themselves to live without endangering their health on the high hills," wrote David-Néel. "It has nothing to do with religion, and so it can be used for ordinary purposes without lack of reverence." David-Néel would rely on the practice over and over again to keep happy, healthy, and heated as she hiked for 19 hours a day in freezing temperatures without food or water, at elevations above 18,000 feet.

"Two more, make them good," says McGee. I can't see him—my eyes are still squinted—but I can hear him, heavy-breathing alongside me, cheering me on. I take another giant inhale, then roll the air up to my chest and exhale, like a wave. I've been doing this for what feels like five minutes. My hands are tingling and my intestines feel like they're slowly uncoiling. I let out an uncontrolled moan.

"Yes!" McGee cheers. "Expression is the opposite of depression! Go for it!"

I moan a little louder, wiggle my body, and breathe a little harder. For a moment, I get self-conscious thinking about the EMTs and the ruddy-faced drunk nearby, who no doubt are watching the spectacle: the middle-aged city boys hyperventilating on a purple BPA-free yoga mat, both of us sounding like dedicated perverts.

This self-expression is an important part of Tummo, McGee said before we started. It reminds me that the stress I'm creating is different from the stress of, say, running late for an important meeting. It is conscious stress. "This is something you are doing to yourself—not something happening to you!" McGee keeps yelling.

The stress that Da Costa's soldiers experienced was unconscious. The men had grown up in rural environments, outside the noise and crowds of the city. The more carnage they saw, the more their unconscious sympathetic responses kept building with no means of release.

Eventually, their nervous systems were so overloaded that they short-circuited and collapsed.

I don't want to short-circuit. I want to condition myself so I can remain flexible to the constant pressures of modern life.

"Keep going," says McGee. "Get it all out!"

Professional surfers, mixed martial arts fighters, and Navy SEALs use Tummo-style breathing to get into the zone before a competition or black ops mission. It's also especially useful for middle-aged people who suffer from lower-grade stress, aches and pains, and slowing metabolisms. For them—for me—Tummo can be a preventative therapy, a way to get a fraying nervous system back on track and keep it there.

Simpler and less intense methods of breathing slow, less, through the nose with a big exhale, can also diffuse stress and restore balance. These techniques can be life-changing, and I'd seen dozens of people changed by them. But they can also take a while, especially for those with long-standing chronic conditions.

Sometimes the body needs more than a soft nudge to get realigned. Sometimes it needs a violent shove. That's what Tummo does.

. . .

That shove is still perplexing to the few scientists paying close attention to such phenomena. They ask: How exactly can conscious extreme breathing hack into the autonomic nervous system?

Dr. Stephen Porges, a scientist and professor of psychiatry at the University of North Carolina, has studied the nervous system and its response to stress for the past 30 years. His primary focus is the vagus nerve, a meandering network within the system that connects to all the major internal organs. The vagus nerve is the power lever; it's what turns organs on and off in response to stress.

When perceived stress level is very high, the vagus nerve slows heart rate, circulation, and organ functions. This is how our reptilian and mammalian ancestors evolved the ability to "play dead" hundreds of millions of years ago, to conserve energy and deflect aggression when under attack by predators. Reptiles still access this ability, as do many mammals. (Imagine a possum or the limp body of a mouse in the jaws of a house cat.)

People "play dead," too, because we share the same mechanisms in the primitive part of our brain stem. We call it fainting. Our tendency to faint is controlled by the vagal system, specifically how sensitive we are to perceived danger. Some people are so anxious and oversensitive that their vagus nerves will cause them to faint at the smallest things, like seeing a spider, hearing bad news, or looking at blood.

Most of us aren't that sensitive. It's much more common, especially in the modern world, to never experience full-blown, life-threatening stress, but to never fully relax either. We'll spend our days half-asleep and nights half-awake, lolling in a gray zone of half-anxiety. When we do, the vagus nerve stays half-stimulated.

During these times, the organs throughout the body won't be "shut down," but will instead be half supported in a state of suspended animation: blood flow will decrease and communication between the organs and the brain will become choppy, like a conversation through a staticky phone line. Our bodies can persist like this for a while; they can keep us alive, but they can't keep us healthy.

Porges found that patients who suffer Da Costa–like maladies such as tingling in their fingers, chronic diarrhea, rapid heart rate, diabetes, and erectile dysfunction are often treated for each of these symptoms with a focus on individual organs. But there's nothing wrong with their stomachs, hearts, or genitals. What they often suffer from are communication problems along the vagal and autonomic network, brought on by chronic stress. To some researchers, it's no coincidence that eight of the top ten most common cancers affect organs cut off from normal blood flow during extended states of stress.

Fixing the autonomic nervous system can effectively cure or lessen these symptoms. In the past decade, surgeons have implanted electrical nodes in patients that work as an artificial vagal nerve to restart blood flow and communication between organs. The procedure is called vagus nerve stimulation, and it's highly effective for patients suffering from anxiety, depression, and autoimmune diseases.

But there is another, less invasive way Porges found to stimulate the vagus nerve: breathing.

Breathing is an autonomic function we can consciously control. While we can't simply decide when to slow or speed up our heart or digestion, or to move blood from one organ to another, we can choose how and when to breathe. Willing ourselves to breathe slowly will open up communication along the vagal network and relax us into a parasympathetic state.

Breathing really fast and heavy on purpose flips the vagal response the other way, shoving us into a stressed state. It teaches us to consciously access the autonomic nervous system and control it, to turn on heavy stress specifically so that we can turn it off and spend the rest of our days and nights relaxing and restoring, feeding and breeding.

"You are *not* the passenger," McGee keeps yelling at me. "You are the pilot!"

This was supposed to be biologically impossible. The autonomic nervous system, per its definition, was supposed to be autonomic, as in automatic, as in beyond our control. And for the past hundred years or so, this belief has held. In much of medicine, it still holds today.

When Alexandra David-Néel finally returned to Paris and wrote about Tummo and other Buddhist breathing techniques and meditations in her 1927 book, *My Journey to Lhasa*, few doctors and medical researchers believed the stories. Few could accept that breathing alone could keep a body warm in freezing temperatures. Fewer believed it could control immune function and heal diseases.

Through the twentieth century, interest in Tummo grew, and a flood of anthropologists, researchers, and seekers traveled to the Himalayas and came back reporting the same feats that David-Néel had been talking about. They told stories of monks wearing nothing but a single layer of clothes throughout the winter, heating themselves in frigid stone monasteries by day and melting circles in the snow around their bare bodies by night. Eventually, a Harvard Medical School researcher named Herbert Benson thought it might be time to put Tummo to the test.

Benson flew to the Himalayas in 1981, recruited three monks, hooked them up to sensors that measured the temperature in their fingers and toes, and then asked them to practice Tummo breathing. During the practice, the temperature in the monks' extremities went up by as much as 17 degrees Fahrenheit and stayed there. The results were published the next year in the esteemed scientific journal *Nature*.

The videos and photographs taken during the Harvard experiments showed short men with satchels wrapped around flabby waists, their skin covered in a thick sheen of sweat, eyes half-closed and lost in a thousand-mile stare. The experiments added credence to what David-Néel and Naropa had described, and yet Benson's monks seemed even stranger than an anarchist opera singer or ancient mystic. It all seemed totally inaccessible to Westerners.

That would change by the early 2000s, when a Dutch man named Wim Hof ran a half-marathon through the snow above the Arctic Circle shirtless and in bare feet. Here was a Westerner who had a beard, thinning lead-colored hair, and a face pulled from a Bruegel painting. In short, he looked like every other middle-aged Northern European male. Hof hadn't grown up in a cave in India or suffered from tuberculosis in a village hospital. He'd worked as a mail carrier and was a father of four.

Years earlier, Hof's wife had taken her own life after years of depression. He had sought refuge from his pain by deepening his practice of yoga, meditation, and breathing practices. He unearthed the ancient technique of Tummo, honed it, simplified it, repackaged it for mass consumption, and began promoting its powers in a string of daredevil stunts that would have been quickly discounted if the media hadn't been around to verify them.

Hof submerged himself in a bath filled with ice for an hour and 52 minutes, and he suffered no hypothermia or frostbite. Then he ran a full marathon in the Namib desert in temperatures that reached 104, without ever sipping a drop of water.

Over the span of a decade, Hof broke 26 world records, each more baffling than the last. These stunts earned him international fame, and his smiling, frost-covered face soon appeared on dozens of magazine covers, in flashy documentary specials, and in a handful of books.

"Wim violated the rules laid out in medical textbooks so drastically scientists had to pay attention," said Andrew Huberman, a professor of neurobiology at Stanford University. Scientists paid attention.

In 2011, researchers at Radboud University Medical Center in the Netherlands brought Hof into a laboratory and started poking and prodding him, trying to figure out how he did what he did. At one point, they injected his arm with an endotoxin, a component of *E. coli*. Exposure to the bacteria usually induces vomiting, headaches, fever, and other flu-like symptoms. Hof took the *E. coli* into his veins and then breathed a few dozen Tummo breaths, willing his body to fight it off. He showed no sign of fever, no nausea. A few minutes later, he rose from the chair and got a cup of coffee.

Hof insisted he wasn't special; neither were David-Néel or the Tibetan monks. Almost anyone could do what they all did. As Hof put it, we just had to "Breathe, motherfucker!"

He proved his point three years later, when Radboud University researchers brought in two dozen healthy male volunteers and randomly split them into two groups. Half the men spent the next ten days learning Hof's version of Tummo while exposing themselves to cold, doing things like playing soccer shirtless in snow. The control group received no training. The two groups were brought back into the lab. Each was hooked up to monitors, then injected with the *E. coli* endotoxin.

The group trained by Hof were able to control their heart rate, temperature, and immune response, and stimulate the sympathetic system. This practice of heavy breathing along with regular cold exposure was later discovered to release the stress hormones adrenaline, cortisol, and norepinephrine on command. The burst of adrenaline gave heavy breathers energy and released a battery of immune cells programmed to heal wounds, fight off pathogens and infection. The huge spike in cortisol helped downgrade short-term inflammatory immune responses, while a squirt of norepinephrine redirected blood flow from the skin, stomach, and reproductive organs to muscles, the brain, and other areas essential in stressful situations.

Tummo heated the body and opened up the brain's pharmacy, flooding the bloodstream with self-produced opioids, dopamine, and serotonin. All that, with just a few hundred quick and heavy breaths.

.   .   .

"One more," says McGee. "Then let it all out and hold."

I do as instructed, and listen as the rushing wind that was pouring through my lungs suddenly stops and is replaced by pure silence, the kind of jarring quietude a skydiver feels the moment a parachute opens. But this stillness is coming from inside. As I hold my breath longer, I feel a comforting heat spread across my body and face. I focus on my heart, rock to its vibrations. Each thump sounds and feels like the kick drum from the beginning of Black Sabbath's "Iron Man."

"Make the silence between your heartbeats last an eternity," McGee says in a soothing voice.

After a minute or so, McGee directs me to take in a huge breath without exhaling, and to hold it again for 15 seconds, gently moving the air around my chest. At his order, I exhale and the cycle starts all over again "Three more rounds," McGee says, his voice raising to a yell. "Be your own superpower!"

As I'm huffing again, I slide my focus to McGee, my cheerleader. He'd told me earlier how he'd been suddenly diagnosed with type 1 diabetes six years ago, at 33. His pancreas shut down and no longer produced insulin. Then he suffered chronic back pain, making him anxious and severely depressed. His blood pressure shot up.

McGee's doctor gave him insulin injections to help stabilize his blood sugar, enalapril to lower his blood pressure, and Valium to ease the pain. "I was also taking four or five ibuprofen every day," he said. But nothing really helped. He only got sicker.

McGee was like 15 percent of the American population—more than 50 million people—who suffer from an autoimmune disorder. In simple terms, these diseases are the result of an immune system that goes rogue and starts attacking healthy tissues. Joints become inflamed, muscles and nerve fibers waste away, rashes cover the skin. These ailments go by many names: rheumatoid arthritis, multiple sclerosis, Hashimoto's disease, type 1 diabetes.

Pharmaceutical treatments, such as immunosuppressants, work by easing symptoms and keeping the patient more comfortable, but they do nothing to address the core malfunction in the body. Autoimmune diseases have no known cure, and even the causes are debated. An increasing body of research has shown that many are tied to dysfunction of the autonomic nervous system.

McGee's awareness of alternative treatments began when a friend mentioned a short feature on someone named "the Ice Man" on Vice TV, the news and culture network. That night McGee tried Wim Hof's heavy breathing technique. "For the first time in a long time, I slept peacefully," he told me. He signed up for Hof's ten-week video course, and within weeks watched as his insulin levels normalized, pain subsided, and blood pressure plunged. He quit taking enalapril and reduced his insulin intake by around 80 percent. He still took ibuprofen, but only a pill or two once a week.

McGee was hooked. He flew to Poland to attend an instructor retreat with Hof, where he and a dozen other students spent two weeks hiking up snowy mountains and swimming in freezing lakes. They breathed a lot. It never felt like a competition, McGee told me, or like some extreme fitness regimen. "Fight it. No pain, no gain. That's all bullshit. That's how you get hurt," McGee explained. The point was to rebalance the body so that it could do what it is naturally adapted to do.

I'd heard dozens of these stories. Men, mainly in their 20s, who'd suddenly been diagnosed with arthritis and psoriasis or depression, who, weeks after practicing heavy breathing, no longer suffered any symptoms. Twenty thousand others in Hof's community exchange blood work data and other metrics of their transformations online. The before-and-after results confirmed their claims. Some of these people were reducing inflammatory markers (C-reactive protein) 40-fold within just a few weeks.

"Doctors say this is more pseudoscience than science, that there's no way any of this can be true," McGee told me. And yet McGee and thousands of other heavy breathers kept showing profound improvements. They kept getting off medications they'd been on for years. They kept heating and healing themselves.

"You cannot copyright breathing, that's part of it, and you can't fault someone for the way they've learned," said McGee. "All you can do is give information."

. . .

Here's the information: To practice Wim Hof's breathing method, start by finding a quiet place and lying flat on your back with a pillow under your head. Relax the shoulders, chest, and legs. Take a very deep breath into the pit of your stomach and let it back out just as quickly. Keep breathing this way for 30 cycles. If possible, breathe through the nose; if the nose feels obstructed, try pursed lips. Each breath should look like a wave, with the inhale inflating the stomach, then the chest. You should exhale all the air out in the same order.

At the end of 30 breaths, exhale to the natural conclusion, leaving about a quarter of the air left in the lungs, then hold that breath for as long as possible. Once you've reached your breathhold limit, take one huge inhale and hold it another 15 seconds. Very gently, move that fresh breath of air around the chest and to the shoulders, then exhale and start the heavy breathing again. Repeat the whole pattern three or four rounds and add in some cold exposure (cold shower, ice bath, naked snow angels) a few times a week.

This flip-flopping—breathing all-out, then not at all, getting really cold and then hot again—is the key to Tummo's magic. It forces the body into high stress one minute, a state of extreme relaxation the next. Carbon dioxide levels in the blood crash, then they build back up. Tissues become oxygen deficient and then flooded again. The body becomes more adaptable and flexible and learns that all these physiological responses can come under our control. Conscious heavy breathing, McGee told me, allows us to bend so that we don't get broken.

. . .

Back on the park lawn, there is no more huffing, no more irritable heart. The journey into self-inflicted sympathetic stress is over. Outside, the world seems to yawn awake in a Disney-like montage: the crackle of pine needles beneath the feet of a squirrel, a brush of wind through the branches, the caw of a distant hawk, all of it broadcast in high fidelity.

Getting here took some effort, and if I wasn't laid out on a mat in a park, breathing this hard for this long could be dangerous. McGee repeatedly told me, as he told all his students, to never, ever practice Tummo while driving, walking, or in "any other environment where you might get hurt if you pass out." And never practice it if you might have a heart condition or are pregnant.

Nobody knows how eliciting such extreme stress might affect the immune and nervous systems in the long term. Some pulmonauts, like Anders Olsson and other slow-and-less proponents, argue that this kind of forced overbreathing could actually be more damaging than it's worth "given the adrenaline society we live in," Olsson told me.

I'm less certain. Alexandra David-Néel used Tummo and other ancient breathing and meditation practices until she died in 1969, at the age of 100. One of her acolytes, a man named Maurice Daubard, is still alive. Daubard had spent his teenage years bedridden in a village hospital with tuberculosis, chronic lung inflammation, and other illnesses. By his 20s, the doctors had given up. Daubard decided to heal himself. He read books, trained in yoga, and taught himself Tummo. He not only completely cured his body of any sickness but gained a superhuman strength.

On his off hours of working as a hairdresser he'd strip to his underwear and run barefoot through snowy forests. Decades before Wim Hof, he immersed himself in ice from the neck down and sat there motionless for 55 minutes. Later, he ran 150 miles beneath the searing sun of the Sahara desert. At 71, he toured the Himalayas on his bike at an elevation of 16,500 feet.

But his greatest feat, Daubard said, was helping thousands of others with illnesses learn the power of Tummo to heal themselves, just as he'd done.

"The human is not only an organism . . . it is also a mind whose strength used wisely can allow us to repair our body when it wobbles," wrote Daubard. As of this writing, Daubard had just turned 89. He still plays harp, reads without glasses, and leads Tummo retreats in the Italian Alps above Aosta, where students join him in stripping down to underwear and sitting in the snow for an hour, then hike half-naked up mountains, and finish with a dip in an ice-covered alpine lake.

"[Tummo] is for the reconstitution of man's immune system," Daubard proclaimed. "It's a fabulous way for the future of man's health."

Tummo isn't the only heavy breathing technique to have recently had a resurgence in the West.

Several years ago, when I was early in my research, I'd heard about a practice called Holotropic Breathwork created by a Czech psychiatrist named Stanislav Grof. The main focus wasn't to reboot the autonomic nervous system or heal the body; it was to rewire the mind. An estimated one million people had tried it, and today more than a thousand trained facilitators run workshops around the world.

I paid a visit to Grof, whose home was just a half hour north of me in Marin County. I drove along a tree-lined street where oak roots the size of human thighs buckled narrow sidewalks and pulled up to the driveway of a Mid-century modern house, grabbed my bag, and approached the front door.

Grof greeted me in a blue oxford shirt, khakis, and clogs. He walked me into his living room past Buddhist figures, Hindu gods, Indonesian masks, and stacks of the 20 books he has written over the years. Two sliding glass doors offered a view of hills sprinkled with red-tile

Spanish-style rooftops. We sat at a redwood patio table and Grof told me how it all began.

It was November 1956, and Grof was a student at the Czechoslovak Academy of Sciences in Prague. The university's psychology department had been sent a sample of a new drug from Sandoz, a Swiss pharmaceutical company. The drug was originally developed to treat menstrual and headache pain, but Sandoz found that the side effects, which included hallucinations, were too severe to make it marketable. Sandoz thought psychiatrists might use it to better understand and communicate with their schizophrenic patients.

Grof volunteered to try it. An assistant strapped him to a chair and injected him with a hundred micrograms. "I saw light as I had never seen it before, I could not believe it existed," Grof later recalled. "My first thought was that I was looking at Hiroshima. I then saw myself above the clinic, Prague, the planet. Consciousness had no boundaries, I was beyond the planet. I had cosmic consciousness."

Grof was one of the first-ever test subjects of lysergic acid diethylamide-25, better known as LSD.

The experience would guide Grof's research at the Czechoslovak Academy of Sciences and, later, at Johns Hopkins University, where he would research psychotherapy treatments with patients. By 1968, the U.S. government had outlawed the use of LSD, so Grof and his wife, Christina, sought a therapy with the same hallucinatory and healing effects that wouldn't get them thrown in jail. They discovered heavy breathing.

The Grofs' technique was essentially Tummo cranked up to 11. It involved lying on a floor in a dark room, with loud music playing, breathing as hard and quickly as you could for up to three hours. Willingly breathing to the point of exhaustion, they found, could place patients in a state of stress where they could access subconscious and unconscious thoughts. Essentially, the therapy helped people blow a fuse in their minds so they could return to a state of groovy calm.

The Grofs called it Holotropic Breathwork, from the Greek *holos*, which means "whole," and *trepein*, which translates to "progressing toward something." Holotropic Breathwork broke the mind down and moved it toward wholeness.

It took some doing. Holotropic Breathwork often included a journey through "the dark night of the soul," where patients would experience a "painful confrontation" with themselves. Sometimes patients might vomit or suffer nervous breakdowns. If they got through all that, mystical visions, spiritual awakenings, psychological breakthroughs, out-of-body experiences, and, sometimes, what Grof called a "mini life-death-rebirth" could follow. It was so powerful that patients reported seeing their whole lives flash before their eyes. It quickly gained popularity with psychiatrists.

"We took psychotic people, people nobody else wanted to deal with, people for whom any medicines weren't working," said Dr. James Eyerman, a psychiatrist who has used the therapy in his practice for the past 30 years.

From 1989 to 2001, Eyerman led more than 11,000 patients at Saint Anthony's Medical Center in St. Louis through Holotropic Breathwork. He documented the experiences of 482 manic-depressives, schizophrenics, and others, and found that the therapy had significant and lasting benefits. A 14-year-old patient who'd tried to slit his own throat breathed a few Holotropic breaths and sailed off into an altered state of "pure consciousness." A 31-year-old woman addicted to several drugs had an out-of-body experience and, afterward, sobered up and went on to lead a 12-step program. Eyerman saw thousands of similar transformations and reported no adverse reactions or side effects. "These patients would get pretty wild, but it worked for them," he told me. "It worked incredibly well. And the hospital staff just couldn't figure out why."

A few smaller studies followed and showed positive outcomes for people with anxieties, low self-esteem, asthma, and "interpersonal problems." But for most of its 50-year history, Holotropic Breathwork has been sparsely studied, and the studies that do exist gauge subjective experience—that is, how people say they felt before and after.

I wanted to feel it for myself, so I signed up for a session.

. . .

On a crisp fall day I drove a few hours north of Grof's home to a hot springs resort tucked beneath the shade of ancient redwoods. There were dusty yurts, heavily bearded men in toe shoes, women in braids wearing turquoise, homemade granola in Mason jars. It was exactly the kind of scene I'd expected. What I didn't expect were the corporate lawyers, architects in pressed polo shirts, and muscle men in military-style flattops who had also gathered here.

A dozen of us walked into an activity room of a dormitory. Half the group lay down on the floor and prepared to breathe while the other half, the sitters, watched over them. I volunteered to be a sitter for a man named Kerry, who wore Armani glasses and asked me not to touch him during the session because he feared any contact might burn his skin.

The music started, a predictable mix of thumping techno with reverberating lutes and Arabic maqam yodels. What happened next was predictable, too. The business folks breathed heavily and wiggled around on their mats but mostly kept calm and to themselves. Meanwhile, the natural healers in the group went apeshit.

After just a few minutes of breathing, a big man named Ben, who lived off the grid in a cabin a few miles up the mountain, sat up and stared in awe at the palms of his hands as if he were holding a magic Hobbit stone. A few more breaths and Ben began snorting and scratching his crotch. He growled and howled like a wolf, then took off around the room on all fours. The therapists running the session snuck up behind Ben and wrestled him to the floor. They sat on him until he transmogrified back into a human.

Behind Ben, a woman named Mary jabbed her eyes with her knuckles and screamed for her mother. "I want my mommy. *I hate you, Mommy.* I want my mommy. *I hate you, Mommy,*" she sobbed in an alternating devil-and-baby voice. She wormed to a corner and curled up like an abused dog. This went on for two hours.

I couldn't help but notice that neither Mary nor Ben were breathing any faster or deeper than anyone else; they weren't breathing any faster than me, and I was just sitting there calmly watching this scene unravel.

By afternoon, the group switched positions and it was my turn to walk through the dark night of the soul. I'll admit, I was pretty dubious at this point, but I gave it my all, breathing as hard as I could for as long as I could. I felt really hot and sweated, then felt very cold and sweated. My legs went numb, and my fingers coiled, a side effect of hyperventilation called tetany. I was convinced I'd entered into some waking dream state, where the sounds, music, and sensations of my surroundings mingled freely with subconscious thoughts and images.

Sometime later, the hollow electric drums, fake cymbal crashes, and keyboard lutes faded back into my consciousness, and it was over. The group was invited to sit around a table and draw mandalas with crayons on what they'd just experienced. I walked outside into the perfumed evening air and drank a warm beer alone in the passenger seat of my car.

On one hand, Holotropic Breathwork was transformative for Ben and Mary, and hundreds of thousands of others who'd experienced it.

On the other, there was obviously some psychosomatic influence going on there. I couldn't help wondering how much of its curative effects were the result of the environment, of the "set and setting," and how much might be a measurable, physical response to breathing so heavily for so long.

Grof believed that at least some visual and introspective experiences were triggered by having less oxygen in the brain.

During rest, about 750 milliliters of blood—enough to fill a full wine bottle—flows through the brain every minute. Blood flow can increase a little during exercise just as it does in other parts of the body, but it will usually stay consistent.

That changes when we breathe heavily. Whenever the body is forced to take in more air than it needs, we'll exhale too much carbon dioxide, which will narrow the blood vessels and decrease circulation, especially in the brain. With just a few minutes, or even seconds, of overbreathing, brain blood flow can decrease by 40 percent, an incredible amount.

The areas most affected by this are the brain's hippocampus and frontal, occipital, and parieto-occipital cortices, which, together, govern functions such as visual processing, body sensory information, memory, the experience of time, and the sense of self. Disturbances in these areas can elicit powerful hallucinations, which include out-of-body experiences and waking dreams. If we keep breathing a little faster and deeper, more blood will drain from the brain, and the visual and auditory hallucinations will become more profound.

In addition, the sustained pH imbalance in the blood sends distress signals throughout the body, specifically to the limbic system, which controls emotions, arousal, and other instincts. Consciously sustaining these stress signals long enough may trick the more primitive limbic system into thinking the body is dying. This could explain why so many people experience sensations of death and rebirth during Holotropic Breathwork. They have consciously driven themselves into a state their brains perceive as potentially lethal, and then lulled themselves back out by conscious breathing.

Grof admitted that researchers were a long way from really understanding the full picture. He was OK with that; he just knew Holotropic Breathwork offered a heavy shove that so many patients needed but weren't getting with other therapies. Heavy breathing alone did for them what nothing else could do.

Nine

HOLD IT

In 1968, Dr. Arthur Kling left his office at the University of Illinois Medical College and caught a flight to Cayo Santiago, a wild and unpopulated island on Puerto Rico's southeast coast. He grabbed some traps, captured a group of wild monkeys, then took the animals back to the lab to conduct a bizarre and cruel experiment. Kling began by opening the monkey's skulls and removed a dollop of brain from each side. He let the monkeys recover, then released them back into the jungle.

Beyond a few scars on their heads, the monkeys looked normal, but something was wrong inside their brains. They had trouble navigating the world. Some starved to death. Others drowned. A few were quickly devoured by other animals. Within two weeks, all of Kling's monkeys were dead.

A couple years later, Kling traveled to Zambia, just upstream from

Victoria Falls, and repeated the experiment. Within seven hours of re-leasing the altered monkeys back into the wild, all had disappeared.

The monkeys all died because they couldn't recognize which animals were prey and which were predators. They didn't sense the danger of wading into a rushing river, swinging from a thin branch, or approach-ing a rival troop. The animals had no sense of fear, because Kling had removed fear from their brains.

Specifically, Kling had clipped out the monkeys' amygdalae, two almond-size nodes at the center of the temporal lobes. The amygdalae help monkeys, humans, and other high-order vertebrates remember, make decisions, and process emotions. These nodes are also believed to be the alarm circuit of fear, signaling threats and initiating a reaction to fight or run away. Without the amygdalae, Kling wrote, all the monkeys "appeared retarded in their ability to foresee and avoid dangerous con-frontations." Without fear, survival was impossible, or, at minimum, ex-tremely precarious.

Back in the United States, a girl whom psychologists would name S. M. was born around this time with a rare genetic condition called Urbach-Wiethe disease. The condition caused cell mutations and a buildup of fatty material throughout her body, giving her skin a lumpy and puffy appearance and making her voice hoarse. When S. M. was ten, the de-posits had spread into her brain. For reasons nobody understands, the disease left most regions unharmed, but destroyed her amygdalae.

S. M. could see, feel, hear, think, and taste just like anyone else. She had a normal IQ, memory, and perception. But as S. M. entered her late teenage years, her sense of fear diminished. She would approach total strangers, stand a few inches from their faces, and describe her most in-timate sexual secrets, never afraid of embarrassment or rejection. She'd walk outside in a violent thunderstorm to chat with a neighbor, never worrying she'd be battered by debris. She'd eat food if it was around, but wouldn't bother stocking up if the cupboards were bare. S. M. had no fear of growing hungry.

She even lost the ability to recognize fear in the faces of those around her. S. M. could easily register happiness, confusion, or the sadness of friends and family, but didn't have a clue when someone was scared or threatened. Worries, stress, and anxiety all dissolved along with her amygdalae.

One day, when S. M. was in her 40s, a man in a pickup truck pulled up and asked her to go on a date. She got in, and the man drove her to an abandoned barn, threw her on the ground, and tore off her clothes. Suddenly, a dog ran into the barn, and the man became nervous that people might be close behind. He zipped up his pants and dusted himself off. S. M. casually got up and followed the man back to his car. She asked to be driven home.

Dr. Justin Feinstein met S. M. in 2006 while getting a PhD in clinical neuropsychology at the University of Iowa. Feinstein specialized in anxieties, specifically in how to get over them. He knew that fear was the core of all anxieties: a fear of gaining weight led to anorexia; fear of crowds led to agoraphobia; fear of losing control led to panic attacks. Anxieties were an oversensitivity to perceived fear, be it spiders, the opposite sex, confined spaces, whatever. On a neuronal level, anxieties and phobias were caused by overreactive amygdalae.

Researchers had spent two decades studying S. M., trying to understand her condition, and trying to scare her. They showed S. M. films of humans eating excrement, took her to theme-park haunted houses, and put slithering snakes on her arms. Nothing worked.

Determined, Feinstein dug deeper and found a study in which human subjects were administered a single breath of carbon dioxide. Even with a small amount, patients reported feelings of suffocation, as if they'd been forced to hold their breath for several minutes. Their oxygen levels hadn't changed and the subjects knew they were never in danger, but many still suffered debilitating panic attacks that lasted for minutes. This wasn't a reaction to a perceived fear or an external threat; it wasn't psychological. The gas was physically triggering some other mechanism in their brains and bodies.

Feinstein and a group of neurosurgeons, psychologists, and research assistants set up an experiment at a laboratory at the University of Iowa hospital. They brought S. M. in and sat her down at a desk, fitting an inhaler mask over her face, connected to an inhaler bag that contained a few lungfuls of 35 percent carbon dioxide and the rest room air. They explained to S. M. that the carbon dioxide would not damage her body, her tissues and brain would have plenty of oxygen. She would never be in any danger. Hearing this, S. M. looked the way she always looked: bored.

"We weren't expecting anything to happen," Feinstein told me. "Nobody was." A few moments later, Feinstein released the carbon dioxide mix into the mouthpiece. S. M. inhaled.

Right away, her droopy eyes grew wider. Her shoulder muscles tensed, her breathing became labored. She grabbed at the desk. "Help me!" she yelled through the mouthpiece. S. M. lifted an arm and waved it as if she were drowning. "I can't!" she screamed. "I can't breathe!" A researcher yanked the mask off, but it didn't help. S. M. jerked wildly and gasped. A minute or so later, she dropped her arms and returned to breathing slowly and calmly.

A single puff of carbon dioxide did to S. M. what no snakes, horror movies, or thunderstorms could. For the first time in 30 years, she'd felt fear, a full-fledged panic attack. Her amygdalae hadn't grown back. Her brain was the same as it had always been. But some dormant switch had suddenly been flipped.

S. M. refused to inhale carbon dioxide again. Years later, the mere idea of it stressed her out. So Feinstein and his researchers confirmed the results with two German twins who also suffered from Urbach-Wiethe disease. The twins had lost their amygdalae, and neither had felt fear in a decade. A single inhalation of carbon dioxide quickly changed that when both suffered the same debilitating anxiety, panic, and crushing fear as S. M.

The textbooks were wrong. The amygdalae were not the only "alarm circuit of fear." There was another, deeper circuit in our bodies that was generating perhaps a more powerful sense of danger than anything the amygdalae alone could muster. It was shared not only by S. M., the German twins, and the few dozen others with Urbach-Wiethe disease, but by everyone and almost every living thing—all people, animals, even insects and bacteria.

It was the deep fear and crushing anxiety that comes from the feeling of not being able to take another breath.

. . .

Take a sip of air through the nose or mouth. For this exercise, it doesn't matter. Now hold it. In a few moments, you'll feel a slight hunger for more. As this hunger mounts, the mind will race, the lungs will ache. You'll become nervous, paranoid, and irritable. You'll start to panic. All senses will zero in on that miserable, suffocating feeling, and your sole desire will be to take another breath.

The nagging need to breathe is activated from a cluster of neurons called the central chemoreceptors, located at the base of the brain stem. When we're breathing too slowly and carbon dioxide levels rise, the central chemoreceptors monitor these changes and send alarm signals to the brain, telling our lungs to breathe faster and more deeply. When we're breathing too quickly, these chemoreceptors direct the body to breathe more slowly to increase carbon dioxide levels. This is how our bodies determine how fast and often we breathe, not by the amount of oxygen, but by the level of carbon dioxide.

Chemoreception is one of the most fundamental functions of life. When the first aerobic life forms evolved two and a half billion years ago, they had to sense carbon dioxide to avoid it. The chemoreception that developed passed up through bacteria to more complex life. It's what stimulates the suffocating feeling you just felt holding your breath.

As humans evolved, our chemoreception became more plastic, meaning it could flex and shift with changing environments. It's this ability to adapt to different levels of carbon dioxide and oxygen that helped humans colonize altitudes 800 feet below and 16,000 feet above sea level.

Today, chemoreceptor flexibility is part of what distinguishes good athletes from great ones. It's why some elite mountain climbers can summit Everest without supplemental oxygen, and why some freedivers can hold their breath underwater for ten minutes. All these people have trained their chemoreceptors to withstand extreme fluctuations in carbon dioxide without panic.

Physical limits are only half of it. Our mental health relies on chemoreceptor flexibility as well. S. M. and the German twins didn't suffer from a debilitating panic attack and anxiety because of mental illness. They suffered because of a broken line of communication between their chemoreceptors and the rest of their brains.

This may sound very basic: of course we're conditioned to panic when we're denied a breath or think we're about to be. But the scientific reason for that panic—that it can be generated by chemoreceptors and *breathing* instead of by external psychological threats processed by the amygdalae—is profound.

All this suggests that for the past hundred years psychologists may have been treating chronic fears, and all the anxieties that come with them, in the wrong way. Fears weren't just a mental problem, and they couldn't be treated by simply getting patients to think differently. Fears and anxiety had a physical manifestation, too. They could be generated from outside the amygdalae, from within a more ancient part of the reptilian brain.

Eighteen percent of Americans suffer from some form of anxiety or panic, with these numbers rising every year. Perhaps the best step in treating them, and hundreds of millions of others around the world, was by first conditioning the central chemoreceptors and the rest of the brain to become more flexible to carbon dioxide levels. By teaching anxious people the art of holding their breath.

As far back as the first century BCE, inhabitants of what is now India described a system of conscious apnea, which they claimed restored health and ensured long life. The Bhagavad Gita, a Hindu spiritual text written around 2,000 years ago, translated the breathing practice of pranayama to mean "trance induced by stopping all breathing." A few centuries after that, Chinese scholars wrote several volumes detailing the art of breathholding. One text, *A Book on Breath by the Master Great Nothing of Sung-Shan*, offered this advice:

> Lie down every day, pacify your mind, cut off thoughts and block the breath. Close your fists, inhale through your nose, and exhale through your mouth. Do not let the breathing be audible. Let it be most subtle and fine. When the breath is full, block it. The blocking (of the breath) will make the soles of your feet perspire. Count one hundred times "one and two." After blocking the breath to the

extreme, exhale it subtly. Inhale a little more and block (the breath) again. If (you feel) hot, exhale with "Ho." If (you feel) cold, blow the breath out and exhale it with (the sound) "Ch'ui." If you can breathe (like this) and count to one thousand (when blocking), then you will need neither grains nor medicine.

Today, breathholding is often associated with disease. "Don't hold your breath," the adage goes. Denying our bodies a consistent flow of oxygen, we've been told, is bad. For the most part, this is sound advice.

Sleep apnea, a form of chronic unconscious breathholding, is terribly damaging, as most of us know by now, causing or contributing to hypertension, neurological disorders, autoimmune diseases, and more. Breathholding during waking hours is injurious as well, and more widespread.

Up to 80 percent of office workers (according to one estimate) suffer from something called continuous partial attention. We'll scan our email, write something down, check Twitter, and do it all over again, never really focusing on any specific task. In this state of perpetual distraction, breathing becomes shallow and erratic. Sometimes we won't breathe at all for a half minute or longer. The problem is serious enough that the National Institutes of Health has enlisted several researchers, including Dr. David Anderson and Dr. Margaret Chesney, to study its effects over the past decades. Chesney told me that the habit, also known as "email apnea," can contribute to the same maladies as sleep apnea.

How could modern science and ancient practices be so at odds?

Again, it comes down to will. The breathholding that occurs in sleep and constant partial attention is *unconscious*—it's something that happens to our bodies, something that's out of our control. The breathholding practiced by the ancients and revivalists is *conscious*. These are practices we will ourselves to do.

And when we do them properly, I'd heard, they can work wonders.

. . .

It's a muggy Wednesday morning and I'm sitting on a rumpled sofa in Justin Feinstein's office at the Laureate Institute for Brain Research in downtown Tulsa, Oklahoma. Across from me is a window that looks out over a cardboard-colored sky and a paisley landscape of red and orange leaves. Feinstein is seated below it, flipping through a stack of scientific papers on a doublewide desk that has not one inch of vacant space. He's wearing an untucked button-down shirt with the cuffs rolled up, flip-flops, and baggy khakis with crayon stains, compliments of his three-year-old daughter. He looks the way you'd imagine a neuropsychologist to look: brainy with a touch of funk.

Feinstein has just been awarded a five-year NIH grant to test the use of inhaled carbon dioxide on patients with panic and anxiety disorders. After his experience administering the gas to S. M. and the German twins with Urbach-Wiethe disease, he'd become convinced carbon dioxide could not only cause panic and anxiety, but that it might also help cure it. He believed that breathing heavy doses of carbon dioxide might elicit the same physical and psychological benefits as the thousand-year-old breathholding techniques.

But his therapy didn't require patients to actually hold their breath or block their throats and count to one hundred with clenched hands like the ancient Chinese. His patients were far too anxious and impatient to practice such an intense technique. Carbon dioxide did all that for them. They'd come in, think about whatever they wanted to think about, take a few inhales of the gas, flex their chemoreceptors back to normal, and be on their way. It was the ancient art of breathholding for those people too anxious to hold their breath.

Breathholding hacks, or, as Feinstein would call them, carbon dioxide therapies, have been around for thousands of years. The ancient Romans prescribed soaking in thermal baths (which contained high levels of carbon dioxide that was absorbed through the skin) as a cure for anything from gout to war wounds. Centuries later, Belle Époque French gathered at thermal springs at Royat in the French Alps to wade in bubbling waters for days at a time.

"The study of the chemical composition of the four mineral springs at Royat will show that we have several powerful agents at our command, and that much is available for the treatment of many morbid conditions, which resist the usual pharmaceutical applications we make use of in daily practice," wrote George Henry Brandt, a British doctor who visited in the late 1870s. Brandt was talking about skin disorders like eczema and psoriasis, along with respiratory maladies like asthma and bronchitis, all of which were "cured almost with certainty" after a few sessions.*

Royat physicians would eventually bottle up carbon dioxide and administer it as an inhalant. The therapy was so effective that it made it stateside in the early 1900s. A mixture of 5 percent carbon dioxide and the rest oxygen made popular by Yale physiologist Yandell Henderson

---

*Since Brandt's reports, thousands of researchers have tested the effects of carbon dioxide therapies on cardiovascular health, weight loss, and immune function. A quick search of "transdermal carbon dioxide therapy" on PubMed brings up more than 2,500 studies. Most of these studies, I've found, confirm what Royat researchers had discovered a hundred years earlier, and the Greeks thousands of years before them: exposing the body to carbon dioxide, whether in water or through injections or via inhalation, increases oxygen delivery to muscles, organs, brain, and more; it dilates arteries to increase blood flow, helps dissolve more fat, and is a powerful treatment for dozens of ailments. For an extensive history of carbon dioxide research and several more resources, visit www.mrjamesnestor.com/breath.

was used with great success to treat strokes, pneumonia, asthma, and asphyxia in newborn babies. Fire departments in New York, Chicago, and other major cities installed carbon dioxide tanks on their trucks. The gas was credited with saving many lives.

All the while, blends of 30 percent carbon dioxide and 70 percent oxygen became a go-to treatment for anxiety, epilepsy, and even schizophrenia. With a few huffs of the stuff, patients who'd spent months or years in a catatonic state would suddenly come to. They'd open their eyes, look around, and begin calmly talking with doctors and other patients.

"It was a wonderful feeling. It was marvelous. I felt very light and didn't know where I was," one patient reported. "I knew something had happened to me and I wasn't sure what it was."

The patients would stay in this coherent, lucid state for about 30 minutes, until the carbon dioxide wore off. Then, without warning, they'd stop mid-sentence and freeze, staring into space and striking statue-like poses or sometimes collapsing. The patients were sick again. They'd stay that way until the next hit of carbon dioxide.

And then, for reasons nobody quite understands, by the 1950s, a century of scientific research disappeared. Those with skin disorders turned to pills and creams; those with asthma managed symptoms with steroids and bronchodilators. Patients with severe mental disorders were given sedatives.

The drugs never cured schizophrenia or other psychoses, but they didn't provoke out-of-body experiences or feelings of euphoria, either. They numbed patients, and continued to numb them for weeks, months, and years—as long as they kept taking them.

"What's interesting to me is that nobody disproved it," says Feinstein of carbon dioxide therapy. "The data, the science, still holds today."

He tells me how he'd stumbled upon some obscure studies by Joseph Wolpe, a renowned psychiatrist who rediscovered carbon dioxide therapy as a treatment for anxiety and had written an influential paper about it in the 1980s. Wolpe's patients shared stunning and long-lasting improvements after just a few huffs. Donald Klein, another renowned psychiatrist and expert in panic and anxiety, suggested years later that the gas might help reset the chemoreceptors in the brain, allowing patients to breathe normally so they could think normally. Since then, few researchers have studied the treatments. (Feinstein estimates there are about five researching it now.) He just kept wondering if the early researchers were right, if this ancient gas might be a remedy to modern ailments.

"As a psychologist, I think, what are my options, what is the best treatment for these patients?" Feinstein says.

Pills, he tells me, offer a false promise and do little good for most people. Anxiety disorders and depression are the most common mental illnesses in the United States, and about half of us will suffer from one or the other in our lifetime. To help cope, 13 percent of us over the age of 12 will use antidepressants, most often selective serotonin reuptake inhibitors, also known as SSRIs. These drugs have been lifesavers for millions, especially those with severe depression and other serious conditions. But less than half the patients who take them get any benefits.* "I keep asking myself," says Feinstein, "Is this the best we can do?"

Feinstein had explored various non-pharmaceutical therapies; he'd spent a decade learning and teaching mindfulness meditation. A wealth of scientific research shows that meditation can change the structure and function of critical areas of the brain, help relieve anxieties, and boost

---

*A 2019 British study published in *The Lancet* found that depressive symptoms were 5 percent lower after six weeks in a group treated with an SSRI, which offered, in the author's words, "no convincing evidence" of an effect. After 12 weeks, there was a 13 percent reduction, a finding the researchers described as "weak."

focus and compassion. It can work wonders, but few of us will ever reap these rewards, because the vast majority of people who try to meditate will give up and move on. For those with chronic anxieties, the percentages are far worse. "Mindful meditation—as it is typically practiced—is just no longer conducive to the new world we live in," Feinstein explains.

Another option, exposure therapy, is a technique that exposes patients repeatedly to their fears so that they become more accepting of them. It's highly effective but takes a while, usually involving many long sessions over weeks or months. Finding psychologists with that kind of time, and patients with the necessary resources, can be a challenge.

But everyone breathes, and, today, few of us breathe well. Those with the worst anxieties consistently suffer from the worst breathing habits.

People with anorexia or panic or obsessive-compulsive disorders consistently have low carbon dioxide levels and a much greater fear of holding their breath. To avoid another attack, they breathe far too much and eventually become hypersensitized to carbon dioxide and panic if they sense a rise in this gas. They are anxious because they're overbreathing, overbreathing because they're anxious.

Feinstein found some inspiring recent studies by Alicia Meuret, the Southern Methodist University psychologist who helped her patients blunt asthma attacks by slowing their breathing to increase their carbon dioxide. This technique worked for panic attacks, too.

In a randomized controlled trial, she and a group of researchers gave 20 panic sufferers capnometers, which recorded the amount of carbon dioxide in their breath throughout the day. Meuret crunched the data and found that panic, like asthma, is usually preceded by an increase in breathing volume and rate and a decrease in carbon dioxide. To stop the attack before it struck, subjects breathed slower and less, increasing their carbon dioxide. This simple and free technique reversed dizziness, shortness of breath, and feelings of suffocation. It could effectively cure a panic attack before the attack came on. "'Take a deep breath' is not a helpful instruction," Meuret wrote. "Hold your breath" is much better.

We leave Feinstein's office and stroll through a labyrinth of elevators and staircases until we enter through soundproofed double doors. This is Feinstein's lair. Through the door to the right, he and his team conduct research on floatation, a therapy that involves lying in a salt-water pool in a dark, soundless room. Through the door to the left is Feinstein's newest project: a carbon dioxide therapy laboratory. It's a tiny window-less box that looks as if it might have held HVAC equipment at one point. We squeeze into the space like clowns in a phone booth. On a folding desk is the usual array of monitors, computers, wires, EKGs, cap-nometers, and other stuff I've grown accustomed to wearing over the past few years. A beat-up yellow cylinder that looks like a Cold War–era Russian missile sits in the corner. Feinstein tells me it holds 75 pounds of pure carbon dioxide.

For the past few months, as part as his NIH research Feinstein has brought in patients suffering with anxiety and panic to this lab and given them a few hits of carbon dioxide. So far, he tells me, the results have been promising. Sure, the gas elicited a panic attack in most pa-tients, but this is all part of the baptism-by-fire process. After that initial bout of discomfort, many patients report feeling relaxed for hours, even days.

I've decided to throw my chemoreceptors into the ring. I've signed up to see what a few heavy doses of carbon dioxide would do to my own body and brain.

Feinstein sticks a piece of white foamy material with a metal sensor on my middle and ring fingers. This device, called a galvanic skin con-ductance meter, will measure small amounts of sweat released during states of sympathetic stress. On my other hand, a pulse oximeter will record my heart rate and oxygen levels.

The mixture I'll inhale is 35 percent carbon dioxide, and the rest is room air—about the same percentage of carbon dioxide once used to test schizophrenics, sans the oxygen. Feinstein administered this same dose on S. M., who panicked and hated it. He also tried it out on a few patients early on, but they too suffered heavy panic attacks. Some patients were so freaked out they refused to take another hit, so Feinstein now reduces the dose to 15 percent—enough to give the chemoreceptors a good workout, but not enough to keep patients from coming back for more. Since I didn't suffer from panic attacks or chronic anxiety, not yet at least, he offered to crank up my dose to the S. M. level to see what happens.

He calmly explains, for the third time today, that any suffocation I might feel after inhaling the gas is only an illusion, that my oxygen levels will remain unchanged, and that I'll be in no danger. Although he means to calm my fears, the constant disclaimers only make me more, well, anxious.

"You good?" Feinstein says, tightening the Velcro straps on the face mask. I nod, take a few last, sweet inhales of room air, and sink deeper in the chair. We'll begin takeoff in two minutes.

As Feinstein walks over to a computer and futzes with cables and tubes and wires, I'm left to sit, stare at my cuticles, and reminisce a bit. My mind wanders to last year, when I first visited Anders Olsson in Stockholm.

It was just after our interview in the co-working reception hall, and Olsson took me into his office, a little hovel filled with research papers, pamphlets, and face masks. A beat-up carbon dioxide tank stood amid the rubble. Olsson told me that he and a group of DIY pulmonauts had been running their own experiments with carbon dioxide over the past couple of years. They weren't interested in the megadoses used to treat epilepsy and mental disorders. Olsson and his crew weren't sick. They were interested in exploring the preventative and performance benefits of the gas, in flexing their chemoreceptors even wider so that they could push their bodies further.

The most effective and safest blend they found was a few huffs of around 7 percent carbon dioxide mixed with room air. This was the "super endurance" level Buteyko found in the exhaled breath of top athletes. Breathing in this mixture had none of the hallucinogenic or panic-inducing effects. You hardly noticed it, and yet it offered potent results. Olsson shared some reports from pulmonauts in the field.

User #1: "So I'm in Toronto now and I decided to go for a rollerblade. I'm a big rollerblader and have done this route by the water on lakeshore many times before. But get this: No matter [how] hard I pushed it, and I pretty much gave it 110% the entire time . . . I not once needed to open my mouth to pant!"

User #2: "I did some carbon dioxide treatments 3 times yesterday, about 15 minutes a piece. And today, I went canoeing and then when I had sex with my girlfriend . . . by the end of it she was panting and tired, and I wasn't even out of breath at all! I felt like I was superhuman!"

User #3: "Holy fuck! . . . I was breathing . . . and I started to feel fricking AMAZing. Euphoric even. To the point where breathing felt automatic."

Olsson hooked up the tank and offered me a few huffs. I felt a slight spaciness, which was soon followed by a slight headache. I was unimpressed.

. . .

Back in Tulsa, Feinstein is about to administer something else entirely. It's several times what I'd had before and several thousand times more than my chemoreceptors are normally exposed to.

He reaches over and points to the big red button on the desk. It switches the air hose from room air to the carbon dioxide in a foil bag hanging on the wall. The bag is a precautionary device. I'll be huffing from it instead of directly out of the tank, in case there's a malfunction in the system, or in my brain. Should a faucet stay open or should I suddenly start panicking uncontrollably, I'll only be able to breathe the contents inside the bag, which works out to about three big huffs.

Next to the red button is a stress dial. It will record my perceived anxiety. It's currently set to 1, the lowest level. When I start feeling anxious after inhaling the gas, I can crank the dial as far up as 20, marking an extreme state of panic.

Over the next 20 minutes, I'll need to take three big inhales of carbon dioxide. I can take all three breaths one after the other if I'm feeling comfortable. If I'm not, I can wait several minutes between hits. The amount of time patients wait provides insight into how intense the experience was.

Strapped in and ready, I'm trying to calm myself, watching the live feed of my vitals on the computer monitor. As I inhale, my heart rate increases, then decreases with every exhale, making a smooth sine wave across the screen. Oxygen hovers at around 98 percent, and exhaled carbon dioxide holds steady at 5.5 percent. All systems are go.

It feels like I'm a fighter pilot on a stealth mission, hissing Darth Vader breaths through a face mask, my hand on a missile-release button. Not the kind of scene I'd ever associated with mental health therapy. But Feinstein's goal isn't to change the way a patient feels on an emotional level. It's to reset the basic mechanics of the primitive brain.

Chemoreceptors, after all, don't care if the carbon dioxide in the bloodstream is generated from strangulation, drowning, panic, or a foil bag on a wall in Tulsa. They set off the same alarm bells. Experiencing such an attack in a controlled environment helps demystify it, teaching patients what an attack feels like before it comes on so we can prevent it. It gives us conscious power over what for too long has been considered an unconscious ailment, and shows us that many of the symptoms we're suffering can be caused, and controlled, by breathing.

One more slow and deep inhale, a thumbs-up, and I close my eyes and push all the air from my lungs. I punch the red button and hear the hose engage to the foil bag, then take in an enormous breath.

The air tastes metallic. It oozes into my mouth, zinging my tongue and gums with the sensation of drinking orange juice from an aluminum cup. The gas pushes deeper, down my throat, coating my innards with what feels like a sheet of aluminum foil. It cracks through the bronchioles, into the alveoli, and into the bloodstream. I brace myself for the hit.

One second. Two seconds. Three. Nothing. I feel no different than I did a few seconds ago or a few minutes before that. I hold the stress dial at 1.

Feinstein said this might happen. He'd given this heavy dose to a Wim Hof practitioner months earlier, and the man barely felt anything. After so much heavy breathing and breathholding, Feinstein hypothesized this subject had already flexed his chemoreceptors wide open. Meanwhile, I had just come off ten days of forced mouthbreathing followed by ten days of forced nasal breathing. I'd raised my resting carbon dioxide levels by 20 percent. I too have probably flexed my chemoreceptors as far as they could reasonably go.

Amid these thoughts, I feel a slight constriction in my throat. It's subtle. I take in a breath of room air, push out an exhale. This requires some effort. The red button is switched off; I'm no longer breathing any more of the carbon dioxide mixture, but it feels like someone has jammed a sock in my mouth. I try to take another breath, but the sock keeps growing.

OK, now there's a pounding in my temples. I open my eyes to check my levels, but the room is blurry. A few seconds later, I'm viewing the world through what looks like cracked and dirty binoculars. I can't breathe. Every sense feels as if it's being torn from my control, vacuumed out.

Maybe 10 or 20 seconds pass before the sock shrinks, there's a cooling at the back of my neck, and the whirlpool of anxiety reverses and floats off. The color and clarity of my vision ripples outward, like a hand clearing mist from a window. Feinstein stands a few feet from me, staring. It all comes back to life. I can breathe again.

I sit there for a few minutes sweating, kind of laughing, kind of crying. I'm trying to prepare myself for two more inhales of this ghastly mixture of gas over the next 15 minutes. Any self-talk I can muster—*This choking is just an illusion; relax, it will only last a few minutes*—does nothing.

After all, the fear I had just felt and would feel again with the next hit won't be mental. It's mechanical; and conditioning the chemoreceptors to widen takes a few sessions, which is why Feinstein's patients come back to re-up over the course of a few days. This is, at its core, an exposure therapy. The more I expose myself to this gas, the more resilient I'll be when I'm overloaded.

And so, in the name of research, and for the sake of my own future chemoreceptor flexibility, I push the red button and take two more hits, one after the other.

And I panic, again and again.

Ten

FAST, SLOW, AND NOT AT ALL

Eight hundred thousand commuters make their way down Paulista Avenue every day, and it shows. The lanes are gridlocked with compact cars and rusting scooters, the sidewalks are a rushing river of men in candy-colored dress shirts, women carrying on intense speakerphone conversations, and schoolgirls wearing T-shirts that their parents certainly didn't translate into English: *I Give Zero Fucks*, *PornFreak*, and *I Got Zero Chill in Me*.

Every few blocks, there's a newsstand selling the requisite *Cosmopolitan* and *Playboy*, but also Nietzsche and Trotsky manifestos, collections of Charles Bukowski's dirty poetry, and Volume 1 of Marcel Proust's 1,056-page ramble, *Remembrance of Things Past*. More honks, a screech of wheels, someone yells something to someone, the light turns green, and we all make our way across the vast intersection, deeper into a canyon of mirrored buildings.

I've come here, to downtown São Paulo, Brazil, to meet with a renowned expert on the foundations of yoga, a man named Luíz Sérgio Álvares DeRose. The yoga DeRose studies and teaches is an ancient practice, vastly different from the yoga at neighborhood studios. It was developed before yoga was even called yoga, before it was an aerobic exercise or had spiritual connotations . . . at a time when it was a technology of breathing and thinking.

I've come to meet with DeRose because after all this research, after so many years of reading books and talking to experts, I've still got questions.

First up, I want to know why the body heats up during Tummo and other Breathing+ practices. The heavy dose of stress hormones could blunt the pain of cold, but it can't stop damage to the skin, tissues, and the rest of the body. Nobody knows how Maurice Daubard, Wim Hof, and their followers can sit naked in the snow for hours and not get hypothermia or frostbite.

Even more confounding are the monks of both Bön and Buddhist traditions, who practice a mellower version of Tummo that stimulates the opposite physiological response. These monks don't huff and puff. Instead, they sit cross-legged and breathe slow and less, inducing a state of extreme relaxation and calm, reducing their metabolic rates by as much as 64 percent—the lowest number recorded in laboratory experiments. The monks should be dead, or at least suffering from extreme hypothermia. However, in this very relaxed state, they're able to increase body temperature by double digits and stay steaming hot in sub-zero temperatures for hours.

Another question that's been baffling me is how heavy Breathing+ techniques like Holotropic Breathwork can induce such hypersurreal and hallucinatory effects. After 15 minutes of conscious overbreathing, the brain starts to compensate. In several studies, there appeared to be no oxygen deprivation associated with conscious overbreathing practices after 'an initial hit. All cognitive functions should be normal, but they certainly are not.

Researchers in the United States and Europe have spent decades sticking nodes and probes into people trying to understand the hidden mechanism behind these techniques. But nobody has found it; nobody can explain it.

So I've decided to look backward, to the ancient texts of the Indians for answers. Every technique I've studied and practiced over the past decade, and every technique I've so far described in this book—from Coherent Breathing to Buteyko, Stough's exhalations to breathholding—first appeared in these age-old texts. The scholars who wrote them clearly knew that breathing is more than just ingesting oxygen, expelling carbon dioxide, and coaxing nervous systems. Our breath also contained another invisible energy, more powerful and affecting than any molecule known to Western science.

DeRose supposedly knows all about it. He's written 30 books on the oldest forms of yoga and breathing. He's been decorated with every imaginable laurel in Brazil, including Emeritus Advisor of the Order of Parliamentarians, Grand Officer of the Order of the Noble Knights of São Paulo, Counselor of the Brazilian Academy of Art, Culture, and History, and dozens more. These were decorations normally reserved for great statespeople. DeRose has them all, even something called a Great Necklace of the Order of Merit from the East Indies.

And now, as I cross over from Paulista on to Rua Bela Cintra, he is only a few blocks away.

.   .   .

Open up a book or website or article or Instagram feed on yoga and chances are you'll see the word *prana*, which translates to "life force" or "vital energy." Prana is, basically, an ancient theory of atoms. The concrete in your driveway, clothes on your body, spouse clanking dishes in your kitchen—they're all made of swirling atomic bits. It's energy. It's prana.

The concept of prana was first documented around the same time in India and China, some 3,000 years ago, and became the bedrock of medicine. The Chinese called it *ch'i* and believed the body contained channels that functioned like prana power lines connecting organs and tissues. The Japanese had their own name for prana, *ki*, as did the Greeks (*pneuma*), Hebrews (*ruah*), Iroquois (*orenda*), and so on.

Different names, same premise. The more prana something has, the more alive it is. Should this flow of energy ever become blocked, the body would shut down and sickness would follow. If we lose so much prana that we can't support basic body functions, we die.

Over the millennia, these cultures developed hundreds—*thousands*—of methods to maintain a steady flow of prana. They created acupuncture to open up prana channels and yoga postures to awaken and distribute the energy. Spicy foods contained large doses of prana, which is one of the reasons traditional Indian and Chinese diets are often hot.

But the most powerful technique was to inhale prana: to breathe. Breathing techniques were so fundamental to prana that ch'i and ruah and other ancient terms for energy are synonymous with respiration. When we breathe, we expand our life force. The Chinese called their system of conscious breathing *qigong*: *qi*, meaning "breath," and *gong*, meaning "work," or, put together, breathwork.

Through the past several centuries of medical advancements, Western science never observed prana, or even confirmed that it exists. But in 1970, a group of physicists took a stab at measuring its effects when a man named Swami Rama walked into the Menninger Clinic in Topeka, Kansas, the largest psychiatric training center in the country at the time.

Rama was wearing a flowing white robe, a mala bead necklace, and sandals, and had hair that drooped down past his shoulders. He spoke eleven languages, ate mostly nuts, fruits, and apple juice, and claimed to have hardly any material possessions. "At six-feet-one-inch and 170 pounds, and with a lot of energy for debate and persuasion, he was a formidable figure," wrote a staff member.

By the age of three, Rama was practicing yoga and breathing techniques around his home in northern India. He'd later move to Himalayan monasteries and study secret practices alongside Mahatma Gandhi, Sri Aurobindo, and other Eastern luminaries. In his 20s, he headed west to attend Oxford and other universities, then eventually set off around the world to teach the methods he'd learned from the masters to anyone who cared to listen.

In the spring of 1970, Rama was sitting at a wooden desk in a small, pictureless office at the Menninger Clinic with an EKG over his heart and EEG sensors on his forehead. Dr. Elmer Green stood over him inspecting the equipment through Coke-bottle glasses. A former Navy weapons physicist, Green headed the Voluntary Controls Program, a lab within the clinic that investigated something called "psychophysiologic self-regulation," or what would become known as the mind/body connection. Green had heard about the extraordinary abilities of Indian meditators from his colleagues and had seen data from a recent experiment with Rama at a Veterans Administration hospital in Minnesota. Green wanted to confirm the results with the latest scientific instruments; he wanted to observe the power of prana for himself.

Rama exhaled, calmed himself, lowered his thick eyelids, and then began breathing, carefully controlling the air entering and exiting his body. The lines on the EEG readout grew longer and softer, from hyperactive beta waves to calming and meditative alpha, and then to long and low delta, the brain waves identified with deep sleep. Rama stayed in this comatose state for a half hour, becoming so relaxed at one point that he began gently snoring. When he "woke up," he gave a detailed recap of the conversation in the room that had occurred while he displayed brain waves of deep sleep. Rama didn't call it deep sleep, though. He called it "yogic sleep," a state in which the mind was active while the "brain slept."

In the next experiment, Rama shifted his focus from his brain to his heart. He sat motionless, breathed a few times, and then, when given a signal, slowed his heart rate from 74 to 52 beats in less than 60 seconds. Later, he increased his heart rate from 60 to 82 beats within eight seconds. At one point, Rama's heart rate went to zero, and stayed there for 17 seconds. Green thought Rama had shut off his heart completely, but upon closer inspection of the EKG, he found that Rama had commanded it to beat at 300 beats per minute.

Blood can't move through the chambers when the heart beats this fast. Swami Rama should have been in cardiac arrest, or dead. But he seemed unaffected. He claimed that he could maintain this state for a half hour. The results of the experiment were later reported in *The New York Times*.

Rama went on to shift the prana (or blood flow, or both) to other parts of his body, willing it from one side of his hand to the other. Within 15 minutes, he was able to create a temperature difference of 11 degrees between his little finger and thumb. Rama's hands never moved.

Oxygen, carbon dioxide, pH levels, and stress hormones played no part in Rama's abilities. As far as is known, his blood gases and nervous system were normal throughout each of the experiments. There was some other strange prana force at play, some more subtle energy Rama had harnessed. Dr. Green and the Menninger team knew it was there; they'd measured its effects on Rama's body and brain. They just had no way to calculate it with any of their machines.

By the early 1970s, Swami Rama had become a bona fide breathing superstar, with his bushy eyebrows and laser-beam eyes showing up in *Time*, *Playboy*, *Esquire*, and, later, on daytime television talk shows like *Donahue*. Nobody in the Western world had seen anything like him before. But it turned out Rama wasn't that special.

A French cardiologist named Thérèse Brosse had recorded a yogi doing the same thing Rama did forty years earlier: stopping and starting his heart on demand. A researcher named M. A. Wenger, at the University of California, Los Angeles, repeated the tests and found yogis who could control not only the beat and pulse strength of their hearts but also the flow of sweat on their foreheads and the temperature of their fingertips. Swami Rama's "superhuman" abilities weren't superhuman at all. They'd been standard practice for hundreds of generations of Indian yogis.

Rama revealed some of his secrets of prana control in group lessons and videos. He recommended students begin by harmonizing their breathing, by removing the pause between inhalations and exhalations so that every breath was one line connected with no end. When this practice felt comfortable, he instructed them to lengthen the breaths.

Once a day, they were to lie down, take a brief inhale, and then exhale to a count of 6. As they progressed, they could inhale to a count of 4 and exhale to 8, with the goal of reaching a half-minute exhale after six months of practice. Upon reaching this 30 count, Rama promised his students, they "will not have any toxins and will be disease-free." In an instructional video, gently stroking his own arm, he said, "Your body will look like smooth body, like silk, you see?"

Infusing the body with prana is simple: you just breathe. But controlling this energy and directing it took a while. Rama obviously had learned something much more powerful in the Himalayas, but as far as I could tell in his books and dozens of instructional videos, he never elaborated.

. . .

The best possible explanation I could find of what the "vital substance" of prana might be and how it might work came not from a yogi but from a Hungarian scientist who nearly flunked out of school as a child, shot himself in the arm to get out of serving in World War I, and later earned a Nobel Prize for groundbreaking work on vitamin C.

His name was Albert Szent-Györgyi, and in the 1940s he'd made his way to the United States and would end up heading the National Foundation for Cancer Research, where he spent years investigating the role of cellular respiration. It was there, working in his laboratory in Woods Hole, Massachusetts, that he proposed an explanation for the subtle energy that drives all life and everything else in the universe.

"All living organisms are but leaves on the same tree of life," he wrote. "The various functions of plants and animals and their specialized organs are manifestations of the same living matter."

Szent-Györgyi wanted to understand the process of breathing, but not in the physical or mental sense, or even at the molecular level. He wanted to know how the breath we take into our bodies interacts with our tissues, organs, and muscles on a subatomic level. He wanted to know how life gained energy from air.

Everything around us is composed of molecules, which are composed of atoms, which are composed of subatomic bits called protons (which have a positive charge), neutrons (no charge), and electrons (negative charge). All matter is, at its most basic level, energy. "We can not separate life from living matter," Szent-Györgyi wrote. "Inevitably, studying living matter and its reactions, we study life itself."

What distinguishes inanimate objects like rocks from birds and bees and leaves is the level of energy, or the "excitability" of electrons within those atoms that make up the molecules in matter. The more easily and often electrons can be transferred between molecules, the more "desaturated" matter becomes, the more *alive* it is.

Szent-Györgyi studied the earliest life forms on Earth and deduced that they were all made up of "weak electron acceptors," which meant they couldn't easily take in or release electrons. He argued that this matter had less energy, so it had less chance to evolve. It just stayed there, mucking around without ever doing much, for millions and millions of years.

Eventually, oxygen, the byproduct of that muck, accumulated in the atmosphere. Oxygen was a strong electron acceptor. As new muck evolved to consume oxygen, it attracted and exchanged many more electrons than older, anaerobic life. With this surplus of energy, early life evolved relatively quickly into plants, insects, and everything else. "The living state is such an electronically desaturated state," wrote Szent-Györgyi. "Nature is simple but subtle."

This premise can be applied to life on the planet today. The more oxygen life can consume, the more electron excitability it gains, the more animated it becomes. When living matter is bristling and able to absorb and transfer electrons in a controlled way, it remains healthy. When cells lose the ability to offload and absorb electrons, they begin to break down. "Taking out electrons irreversibly means killing," wrote Szent-Györgyi. This breakdown of electron excitability is what causes metal to rust and leaves to turn brown and die.

Humans "rust" as well. As the cells in our bodies lose the ability to attract oxygen, Szent-Györgyi wrote, electrons within them will slow and stop freely interchanging with other cells, resulting in unregulated and abnormal growth. Tissues will begin "rusting" in much the same way as other materials. But we don't call this "tissue rust." We call it cancer. And this helps explain why cancers develop and thrive in environments of low oxygen.

The best way to keep tissues in the body healthy was to mimic the reactions that evolved in early aerobic life on Earth—specifically, to flood our bodies with a constant presence of that "strong electron acceptor": oxygen. Breathing slow, less, and through the nose balances the levels of respiratory gases in the body and sends the maximum amount of oxygen to the maximum amount of tissues so that our cells have the maximum amount of electron reactivity.

"In every culture and in every medical tradition before ours, healing was accomplished by moving energy," said Szent-Györgyi. The moving energy of electrons allows living things to stay alive and healthy for as long as possible. The names may have changed—prana, orenda, ch'i, ruah—but the principle has remained the same. Szent-Györgyi apparently took that advice. He died in 1986, at the age of 93.

· · ·

A knock, a swing of the door, a few exchanged *bom dias*, and I take a seat in the lobby of the reception area of DeRose's studio complex. There are wood floors and puffy sofas, white walls and framed posters of world maps. A sign at the center of the room reads "Stop and Breathe."

A gaggle of DeRose teachers and students are lounging and giggling in Portuguese at the center of the lobby while they sip chai tea from ceramic cups. Heduan Pinheiro is among them. He's wearing a creaseless shirt and white pants and resembles a 1980s teen sitcom star. Pinheiro has graciously offered to take time away from his busy schedule operating two DeRose Method studios just north of here to be my guide and translator. We walk through the reception area and up a dark staircase to meet the man he calls "The Master."

The small office is decorated with medals and silver swords, each emblazoned with the kinds of Masonic pyramids and eyeballs you see on the backs of dollar bills and on old buildings. "They just give me these things, I don't know why!" says DeRose, vigorously shaking my hand.

He's powerfully built, with a neatly trimmed white beard and wide brown eyes. The shelves behind him are filled with copies of the books he's sold by the millions on pranayama, karma, and other secrets of ancient yoga. I've read a few of them and found no surprises, no secret breathing method I hadn't already known about and tried over the past several years.

That wasn't surprising, either. The history of yoga and the earliest breathing techniques is long established. But now, finally here, I am eager to compare notes with DeRose. I'm eager to see what he knows about prana and the lost art and science of breathing that I don't.

"Shall we begin?" he says.

.   .   .

If you were to travel back in time some 5,000 years to the borders of what is now Afghanistan, Pakistan, and northwestern India, you'd see sand, rocky mountains, dusty trees, red soil, and wide-open plains, the same landscape that now covers most of the Middle East. But you'd also find something else: five million people living in cities of baked-brick tract houses, roads meticulously constructed in geometric patterns, and children playing with copper, bronze, and tin toys. Between the cul-de-sacs, you'd see public bathing pools with running water, and toilets piped to complex sanitation systems. In the marketplace, you'd see tradespeople measuring goods with weights and standardized rulers, sculptors carving elaborate figures into stone, and ceramists throwing pots and tablets.

This was the Indus-Sarasvati civilization, named after the two rivers that flowed through the valley. The Indus-Sarasvati was the largest geographically—some 300,000 square miles—and one of the most advanced of ancient human civilizations. As far as is known, the Indus Valley had no churches or temples or sacred spaces. The people who lived there produced no praying sculptures, no iconography. Palaces, castles, and imposing governmental buildings didn't exist. There was, perhaps, no belief in God.

But the people here believed in the transformative power of breathing. A seal engraving unearthed from the civilization in the 1920s depicts a man in an unmistakable pose. He's sitting erect with his arms outstretched and hands with thumbs in front placed on his knees. His legs are crossed and the soles of his feet are joined, the toes pointing down. His belly is filled with air, as he consciously inhales. Several other unearthed figures share this same posture. These artifacts are the first documented "yogic" postures in human history, which makes sense. The Indus Valley was the birthplace of yoga.

Things seemed to be going so well in the region, until around 2000 BCE, when a drought hit, causing much of the population to disperse. Then Aryans from the northwest moved in. These weren't the blond-haired, blue-eyed soldiers of Nazi lore but black-haired barbarians from Iran. The Aryans took the Indus-Sarasvati culture and codified, condensed, and rewrote it in their native language of Sanskrit. It's from these Sanskrit translations that we get the Vedas, religious and mystical texts that contain the earliest known documentation of the word "yoga." In two texts based on Vedic teachings, the Brihadaranyaka and the Chandogya Upanishads, are the earliest lessons of breathing and the control of prana.

Over the next few thousand years, the ancient breathing methods spread throughout India, China, and beyond. By around 500 BCE, the techniques would be filtered and synthesized into the Yoga Sutras of Patanjali. Slow breathing, breathholding, deep breathing into the diaphragm, and extending exhalations all first appear in this ancient text. A broad interpretation of a passage from Yoga Sutra 2.51 reads:

> When a wave comes, it washes over you and runs up the
> beach. Then, the wave turns around, and recedes over you,

going back to the ocean. . . . This is like the breath, which exhales, transitions, inhales, transitions, and then starts the process again.

There is no mention in the Yoga Sutras of moving between or even repeating poses. The Sanskrit word *asana* originally meant "seat" and "posture." It referred both to the act of sitting and the material you sit on. What it specifically *did not* mean was to stand up and move about. The earliest yoga was a science of holding still and building prana through breathing.

DeRose got a taste for this kind of ancient yoga in the 1970s, when he was traveling through India trying to piece together the earliest practices of the Indus Valley. He'd attended a class in the foothills of the Himalayas, in Rishikesh, India. The studio was very basic, with a dirt floor, and filled with villagers seeking to warm themselves during cold days.

The classes were casual and the relationship between students and teachers was respectful, yet light. Teachers joked with students during exercises; students joked back. "Push yourself!" the instructors would yell in gruff and direct voices. "You can do better than that!" There were no "gymnastics, anti-gymnastics, bioenergetics, occultism, spiritism, zen, dance, body expression, macrobiotics, shiatsu," DeRose would recall. Poses were done once and held for an excruciatingly long time. These long postures allowed students to focus entirely on their breathing. The class was difficult, and at the end of it DeRose was sweaty and sore.

"Nothing like yoga today," he says from across the desk. He tells me that only in the twentieth century would yoga poses be combined and repeated into a kind of aerobic dance called "vinyasa flow." It's this form of yoga and other hybridized techniques that are now taught in gyms, studios, and classrooms. Ancient yoga, and its focus on prana, sitting, and breathing, has turned into a form of aerobic exercise.

That isn't to say modern yoga is bad in any way. It is simply a different practice from the one that first originated 5,000 years ago. An estimated two billion people now practice this modern form because it makes them feel better and look better and stay more flexible in all the ways stretching and exercise does. Hundreds of studies have confirmed the healing benefits of vinyasa flow and asanas, standing up and otherwise.

But what have we lost?

DeRose would spend 20 years flying from Brazil to India, learning Sanskrit and digging up ancient yoga texts "inch by inch, through centuries of debris," he wrote. He found confirmation of the original practices as "Yôga" (pronounced yoooooga), which comes from the ancient Niríshwarasámkhya lineage, a practice and philosophy so different from the modern version that DeRose believes it deserves to be referred to by its ancient name.

Yôga practices were never designed to cure problems, he tells me. They were created for healthy people to climb the next rung of potential: to give them the conscious power to heat themselves on command, expand their consciousness, control their nervous systems and hearts, and live longer and more vibrant lives.

Toward the end of our hours-long meeting I tell DeRose about my experience in that Victorian house ten years ago, how I had practiced an ancient pranayama technique called Sudarshan Kriya and was quickly floored. I tell him how a mellower version of that reaction keeps happening to me, and millions of others, whenever we use traditional yogic breathing.

Versions of *kriya* had been around since 400 BCE, and by some accounts were used by everyone from Krishna to Jesus Christ, Saint John to Patanjali. The kriya I'd experienced was developed in the 1980s by a man named Sri Sri Ravi Shankar and is now practiced by tens of millions of people around the world through The Art of Living Foundation. It does much of what Tummo does because, DeRose says, both were designed from the same ancient practices.*

Sudarshan Kriya, too, was no picnic. It took time, dedication, and will power. The central method, called Purifying Breath, requires more than 40 minutes of intensive breathing, from huffing and at a rate of more than a hundred breaths per minute, to several minutes of slow breathing, and then hardly breathing at all. Rinse and repeat.

I tell DeRose about the extreme sweating, complete loss of time, and lightness I felt for days afterward. How I'd spent the last decade looking for an explanation, conducting various lab experiments, analyzing my blood gases, and scanning my brain.

He sits calmly with his hands neatly folded. He'd heard it all so many times. I'd found nothing in these readouts or scientific measurements, he says, because I've been looking in the wrong place.

It's energy; it's prana. What had happened was simple and common. I'd built up too much prana breathing so heavily for so long, but I hadn't yet adapted to it. This explained the waterworks and shift of consciousness. Sudarshan is derived from two words: *su*, meaning "good," and *darshan*, meaning "vision." In my case, I'd had a very good vision.

Ancient yogis spent thousands of years honing pranayama techniques, specifically to control this energy and distribute it throughout the body to provoke their "good visions," toned down a notch or two. This process should take several months or years to master. Modern breathers like me can try to hack this process and speed it up. But we will fail.

---

*Although Sudarshan and other kriyas may not have been originally created to help sick people get well, they do that anyway. More than 70 independent studies conducted at Harvard Medical School, Columbia College of Physicians and Surgeons, and other institutions found that Sudarshan Kriya was a highly effective treatment for a range of ailments, everything from chronic stress to joint pain and autoimmune diseases.

Hallucinations, howling, soiling one's clothes: none of that was supposed to happen. It's a sign we've gone overboard.

The key to Sudarshan Kriya, Tummo, or any other breathing practice rooted in ancient yoga is to learn to be patient, maintain flexibility, and slowly absorb what breathing has to offer. My initial experience with Sudarshan Kriya may have been a bit jarring, DeRose says, but it also convinced me of the sheer power of breathing.

In the end, it's what brought me here.

After a few more rounds of Q&A with DeRose, it's time for me to go. He needs to pack his things and head back to New York, where his colleagues run two bustling DeRose Method studios in Tribeca and Greenwich Village. I need to catch a 17-hour flight back home.

We exchange a few *obrigados*, shake hands, and I follow Pinheiro, my translator, past the glinting swords and red ribbons into the dark shadows of the hallway. But before I leave, Pinheiro has offered to teach me some of the ancient Yôga breathing techniques DeRose is known for.

We hike up to the third floor, take off our shoes, and step into the studio. The room is no different from any other yoga studio I've seen. There's blue padding on the floor, wall-wide mirrors, bookshelves and Sanskrit posters. Pinheiro kneels cross-legged so that his body is centered between the windows, casting a Buddha shadow across the room. I take a seat across from him. A minute later, we begin to breathe.

We start with jiya pranayama, which involves curling the tongue to the back of the mouth and holding the breath. We run through some bhandas, a method of redirecting and holding prana inside the body by contracting muscles in the throat, abs, and other areas. Then I lie down in front of him so that I'm looking up at the white acoustic tiles on the ceiling. The final exercise I'll be doing, he tells me, is intended to build prana in the body and focus the mind.

"Concentrate on just one fluid movement from inhale to exhale," says Pinheiro. These are the same instructions I'd heard in that Sudarshan Kriya session way back when, the same instructions I learned from Anders Olsson years later, and from Wim Hof Method instructor Chuck McGee. I know this process now; I know these ropes.

I relax my throat and take a very deep inhale into the pit of my stomach, then exhale completely. Inhale again, and repeat.

"All the way in and all the way out," says Pinheiro. "Keep going! Keep breathing!"

· · ·

There it is, again. And here I am, again. That ringing in the ears. The heavy-metal double-bass drum thumping in my chest. The warm static flowing to my shoulders and face. The wave comes, washes over and runs up, then turns around and recedes, back to the ocean.

I've felt this all before now, so many times. It's the same thing the ancient people of the Indus Valley must have experienced 5,000 years ago, and the ancient Chinese 2,000 years after them. Alexandra David-Néel warmed herself with it in a cave in the Himalayas, and Swami Rama focused it on his hands and heart. Buteyko rediscovered it by a window in the asthma ward of the First Moscow Hospital, and Carl Stough taught it to dying veterans at the VA Medical Center in New Jersey.

As I breathe a little faster, go a little deeper, the names of all the techniques I've explored over the past ten years all come back in a rush.

Pranayama. Buteyko. Coherent Breathing. Hypoventilation. Breathing Coordination. Holotropic Breathwork. Adhama. Madhyama. Uttama. Kêvala. Embryonic Breath. Harmonizing Breath. The Breath by the Master Great Nothing. Tummo. Sudarshan Kriya.

The names may have changed over the years, the techniques may have been repurposed and repackaged in different cultures at different times for different reasons, but they were never lost. They've been inside us all this time, just waiting to be tapped.

They give us the means to stretch our lungs and straighten our bodies, boost blood flow, balance our minds and moods, and excite the electrons in our molecules. To sleep better, run faster, swim deeper, live longer, and evolve further.

They offer a mystery and magic of life that unfolds a little more with every new breath we take.

A LAST GASP

Nothing about this place had changed. The threadbare Persian rug. The paint-chipped window that rattles in the breeze. The rumble of diesel trucks chugging up Page Street and the jaundiced streetlight illuminating bits of falling lint. A few of the faces are the same, too: there's the guy with prisoner eyes, the other with Jerry Lewis bangs, and the blond woman with an unplaceable Eastern European accent. I find my usual spot in the corner and sit by the window.

It's been ten years since I came to this room and felt the possibilities of breathing. A decade of traveling, research, and self-experimentation. In that time I've learned that the benefits of breathing are vast, at times unfathomable. But they're also limited.

This became disturbingly clear several months ago. I was in Portland, Oregon, and had just finished a lecture based on the contents of this book. I left the podium and walked to the lobby to talk with a friend when a woman approached. Her eyes were wide, fingers jittery. She told me her mother had just suffered a pulmonary embolism and desperately needed a breathing technique to remove blood clots from her lungs.

A few weeks later, a woman seated next to me on a flight noticed photographs of skulls on my laptop. She asked what I was working on. After I told her, she explained how her friend was suffering from a serious eating disorder, osteoporosis, and cancer. No treatments had worked. She asked if I could please prescribe a breathing practice to bring her friend back to health.

What I explained to each of these people, and what I'd like to make clear now, is that breathing, like any therapy or medication, can't do everything. Breathing fast, slow, or not at all can't make an embolism go away. Breathing through the nose with a big exhale can't reverse the onset of neuromuscular genetic diseases. No breathing can heal stage IV cancer. These severe problems require urgent medical attention.

I wouldn't be alive without antibiotics, immunizations, and a last-minute rush to the doctor's office to zap out a lymph node infection. The medical technologies developed over the last century have saved innumerable lives. They have increased the quality of life around the world many times over.

But modern medicine, still, has its limitations. "I'm dealing with the walking dead," said Dr. Michael Gelb, who'd spent 30 years working as a dental surgeon and sleep specialist. He echoed what I'd heard from Dr. Don Storey, my father-in-law, who'd worked as a pulmonologist for the past 40 years. Dozens of doctors at Harvard, Stanford, and other institutions told me the same thing. Modern medicine, they said, was amazingly efficient at cutting out and stitching up parts of the body in emergencies, but sadly deficient at treating milder, chronic systemic maladies—the asthma, headaches, stress, and autoimmune issues that most of the modern population contends with.

These doctors explained, in so many words and in so many ways, that a middle-aged man complaining of work stress, irritable bowels, depression, and an occasional tingling in his fingers wasn't going to get the same attention as a patient with kidney failure. He'd be prescribed a blood pressure medication and an antidepressant and sent on his way. The role of the modern doctor was to put out fires, not blow away smoke.

Nobody was happy with this arrangement: doctors were frustrated that they had neither the time nor the support to prevent and treat milder chronic problems, while patients were learning that their cases weren't dire enough for the attention they sought.

This is one of the reasons I believe so many people, and so many medical researchers, have come to breathing.

Like all Eastern medicines, breathing techniques are best suited to serve as preventative maintenance, a way to retain balance in the body so that milder problems don't blossom into more serious health issues. Should we lose that balance from time to time, breathing can often bring it back.

"More than sixty years of research on living systems has convinced me that our body is much more nearly perfect than the endless list of ailments suggests," wrote Nobel laureate Albert Szent-Györgyi. "Its shortcomings are due less to its inborn imperfections than to our abusing it."

Szent-Györgyi was talking about sicknesses of our own making, or, as anthropologist Robert Corruccini has called them, "diseases of civilization." Nine out of ten of the top killers, such as diabetes, heart disease, and stroke are caused by the food we eat, water we drink, houses we live in, and offices we work in. They are diseases humanity created.

While some of us may be genetically predisposed toward one disease or another, that doesn't mean we're predestined to get these conditions. Genes can be turned off just as they can be turned on. What switches them are inputs in the environment. Improving diet and exercise and removing toxins and stressors from the home and workplace have a profound and lasting effect on the prevention and treatment of the majority of modern, chronic diseases.

Breathing is a key input. From what I've learned in the past decade, that 30 pounds of air that passes through our lungs every day and that 1.7 pounds of oxygen our cells consume is as important as what we eat or how much we exercise. Breathing is a missing pillar of health.

"If I had to limit my advice on healthier living to just one tip, it would be simply to learn how to breathe better," wrote Andrew Weil, the famed doctor.

Though researchers still have much to learn about this endlessly expansive field, there is plenty of consensus right now about what "breathing better" looks like.

In a nutshell, this is what we've learned.

SHUT YOUR MOUTH

Two months after the Stanford experiment ended, Dr. Jayakar Nayak's lab emailed Anders Olsson and me the results of our 20-day study. The major takeaway we already knew: mouthbreathing is terrible.

After just 240 hours of breathing only through our mouths, catecholamine and stress-related hormones spiked, suggesting that our bodies were under physical and mental duress. A diphtheroid *Corynebacterium* bug had also infested my nose. If I'd continued breathing only through my mouth for a few more days, it might have developed into a full-fledged sinus infection. All the while, my blood pressure was through the roof and my heart rate variability plummeted. Olsson's data mirrored mine.

By night, the constant flow of unpressurized, unfiltered air flowing in and out of our gaping mouths collapsed the soft tissue in our throats to such an extent that we both began to experience persistent nocturnal suffocation. We snored. A few days later, we started choking on ourselves, suffering from bouts of sleep apnea. Had we continued breathing through our mouths, there's a decent chance we both would have developed chronic snoring and obstructive sleep apnea, along with the hypertension and metabolic and cognitive problems that come with it.

Not all of our measurements changed. Blood sugar levels weren't affected. Cell counts in the blood and ionized calcium remained the same, as did most other blood markers.

There were a few surprises. My lactate levels, a measure of anaerobic respiration, actually decreased with mouthbreathing, which suggested I was using more oxygen-burning aerobic energy. This was the opposite of what most fitness experts would have predicted. (Olsson's lactate slightly increased.) I lost about two pounds, due most likely to exhaled water loss. But trust me on this: a post-holiday mouthbreathing diet is not recommended.

The nagging fatigue, irritation, testiness, and anxiety. The horrid breath and constant bathroom breaks. The spaciness, stares, and stomachaches. It was awful.

The human body has evolved to be able to breathe through two channels for a reason. It increases our chances of survival. Should the nose get obstructed, the mouth becomes a backup ventilation system. The few gasping breaths Stephen Curry takes before dunking a basketball, or a sick kid huffs when he has a fever, or you take in when you're laughing with your friends—this temporary mouthbreathing will have no long-term effects on health.

Chronic mouthbreathing is different. The body is not designed to process raw air for hours at a time, day or night. There is nothing normal about it.

BREATHE THROUGH YOUR NOSE

The day Olsson and I removed the plugs and tape, our blood pressure dropped, carbon dioxide levels rose, and heart rates normalized. Snoring decreased significantly from the mouthbreathing phase, from several hours a night to a few minutes. Within two days, neither of us was snoring at all. The bacterial infection in my nose quickly cleared up without treatment. Olsson and I had cured ourselves by breathing through our noses.

Ann Kearney, the doctor of speech-language pathology at the Stanford Voice and Swallowing Center, was so impressed by our data and her own transformation overcoming congestion and mouthbreathing that, at this writing, she is putting together a two-year study with 500 subjects to research the effects of sleep tape on snoring and sleep apnea.

The benefits of nasal breathing extended beyond the bedroom. I increased my performance on the stationary bike by about 10 percent. (Olsson had more modest gains, about 5 percent.) These results paled in comparison to the gains reported by sports training expert John Douillard, but I couldn't imagine any athlete who wouldn't want a 10 percent—or even a 1 percent—advantage over a competitor.

On a more personal note, those first few nasal breaths after ten days of obstruction were so shimmering and rousing that I got a little teary-eyed. I thought about my interviews with all the empty nose syndrome sufferers who'd been told they were crazy, that they should just quit complaining and breathe through their mouths. I thought about kids who'd been told that chronic allergies and congestion were a part of childhood, and the adults who'd convinced themselves that choking every night was a natural part of growing old.

I had felt their pain, and was lucky enough to breathe life on the other side. It's something I'll never forget, and will never, ever repeat.

EXHALE

Carl Stough spent a half century reminding his students of how to get all the air out of our bodies so that we could take more in. He trained his clients to exhale longer and, in the process, do what had long been considered biologically impossible. Emphysemics reported almost total recovery from their incurable conditions, opera singers gained more resonance and tone in their voices, asthmatics no longer suffered from attacks, and Olympic sprinters went on to win gold medals.

As basic as this sounds, full exhalations are seldom practiced. Most of us engage only a small fraction of our total lung capacity with each breath, requiring us to do more and get less. One of the first steps in healthy breathing is to extend these breaths, to move the diaphragm up and down a bit more, and to get air out of us before taking a new one in.

"The difference in breathing in the coordinated pattern and in an altered pattern is the difference between operating at peak efficiency and just getting along," Stough wrote in the 1960s. "An engine does not have to be in tip-top condition to work, but it gives a better performance if it is."

CHEW

The millions of ancient skeletons in the Paris quarries and hundreds of pre–Industrial Age skulls at the Morton Collection had three things in common: huge sinus cavities, strong jaws, and straight teeth. Almost all humans born before 300 years ago shared these traits because they chewed a lot.

The bones in the human face don't stop growing in our 20s, unlike other bones in the body. They can expand and remodel into our 70s, and likely beyond. Which means we can influence the size and shape of our mouths and improve our ability to breathe at virtually any age.

To do this, don't follow the diet advice of eating what our great-grandmothers ate. Too much of that stuff was already soft and overly processed. Your diet should consist of the rougher, rawer, and heartier foods our great-great-great-great-great-great-grandmothers ate. The kinds of foods that required an hour or two a day of hard chewing. And in the meantime, lips together, teeth slightly touching, and tongue on the roof of the mouth.

BREATHE MORE, ON OCCASION

Since meeting Chuck McGee at that roadside park in the Sierras, I've been practicing Tummo with dozens of others from around the world on Monday nights. That's when McGee hosts a free online session open to anyone who wants to "become the eye of the storm."

Overbreathing has gotten a bad rap in the past few decades, and rightfully so. Feeding the body more air than it needs is damaging for the lungs right down to the cellular level. Today, the majority of us breathe more than we should, without realizing it.

Willing yourself to breathe heavily for a short, intense time, however, can be profoundly therapeutic. "It's only through disruption that we can be normal again," McGee told me. That's what techniques like Tummo, Sudarshan Kriya, and vigorous pranayamas do. They stress the body on purpose, snapping it out of its funk so that it can properly function during the other 23½ hours a day. Conscious heavy breathing teaches us to be the pilots of our autonomic nervous systems and our bodies, not the passengers.

HOLD YOUR BREATH

Several months after experimenting with carbon dioxide therapy, I was at home reading the Sunday paper, flipping through the obituaries, and saw that Dr. Donald Klein had died. Klein was the psychiatrist who spent years studying the links between chemoreceptor flexibility, carbon dioxide, and anxieties. He was 90. It was Klein's research that inspired Justin Feinstein to pursue the NIH-funded experiments in Tulsa.

I wrote Feinstein with the news. He was crushed. He told me he'd been planning on reaching out to Klein in the coming weeks regarding what could be a "game-changing discovery."

It turns out that the amygdalae, those gooey nodes on the sides of our head that help govern perceptions of fear and emotions, also control aspects of our breathing. Patients with epilepsy who have had these brain areas stimulated with electrodes immediately cease breathing. The patients were totally unaware of it and didn't seem to feel their carbon dioxide levels rising long after their breathing ceased.

Communication between the chemoreceptors and amygdalae works both ways: these structures are constantly exchanging information and adjusting breathing every second of every minute of the day. If communication breaks down, havoc ensues.

Feinstein believes that people with anxiety likely suffer from connection problems between these areas and could unwittingly be holding their breath throughout the day. Only when the body becomes overwhelmed by carbon dioxide would their chemoreceptors kick in and trigger an emergency signal to the brain to immediately get another breath. The patients would reflexively start fighting to breathe. They'd panic.

This could explain why patients with clinically smaller amygdalas have such a hypersensitivity to carbon dioxide and why they perform so poorly when trying to hold their breath for more than a few seconds.

"What anxious patients could be experiencing is a completely natural reaction—they're reacting to an emergency in their bodies," said Feinstein. "It could be that anxiety, at its root, isn't a psychological problem at all."

This approach is all very theoretical, Feinstein warned, and needs to be rigorously tested, which is what he will do in the coming years. But if it's true, it could explain why so many drugs don't work for panic, anxiety, and other fear-based conditions, and how slow and steady breathing therapy does.

HOW WE BREATHE MATTERS

I've chatted with Anders Olsson every few weeks since we paid through the nose in the Stanford experiment. Our talks are never dull. "I have more energy and focus than ever in my life!" he told me, right after celebrating his 50th birthday. Olsson is a pulmonaut in the purest sense: self-taught and driven by a sense that we are missing something right in front of us, a truth basic and essential.

Through all my travels and travails, there is one lesson, one equation, that I believe is at the root of so much health, happiness, and longevity. I'm a bit embarrassed to say it has taken me a decade to figure this out, and I realize how insignificant it might look on this page. But lest we forget, nature is simple but subtle.

The perfect breath is this: Breathe in for about 5.5 seconds, then exhale for 5.5 seconds. That's 5.5 breaths a minute for a total of about 5.5 liters of air.

You can practice this perfect breathing for a few minutes, or a few hours. There is no such thing as having too much peak efficiency in your body.

Olsson told me he's working on several more devices to help us breathe at this rate—slowly and less. He's finishing production on his BreathIQ, a portable device that measures nitric oxide, carbon dioxide, ammonia, and other chemicals in exhaled breath. Then there are other skunkworks to mimic the effects of perfect breathing: a carbon dioxide suit, a hat, and . . .

Meanwhile, Google just released an app that pops up automatically when the words "breathing exercise" are searched. It trains visitors to inhale and exhale every 5.5 seconds. Down the street from my house is a startup called Spire, which created a device that tracks breath rate and alerts users every time respiration becomes too fast or disjointed. In the fitness industry, resistance masks and mouthpieces with names like Expand-a-Lung are all the rage.

Before we know it, breathing slow, less, and through the nose with a big exhale will be big business, like so much else. But be aware that the stripped-down approach is as good as any. It requires no batteries, Wi-Fi, headgear, or smartphones. It costs nothing, takes little time and effort, and you can do it wherever you are, whenever you need. It's a function our distant ancestors practiced since they crawled out of the sludge two and a half billion years ago, a technology our own species has been perfecting with only our lips, noses, and lungs for hundreds of thousands of years.

Most days, I treat it like a stretch, something I do after a long time sitting or stressing to bring myself back to normal. When I need an extra boost, I come here, to this old Victorian house in the Haight-Ashbury, and sit beside this rattling window with the other Sudarshan Kriya breathers I first met ten years ago.

. . .

The room is packed now, 20 of us sitting in a circle unkinking our necks and pulling fleece blankets onto our laps. The instructor hits the switch on the wall, the lights dim, and long shadows from the street cast across the floor. In the darkness, he thanks us for coming, brushes his bangs aside, adjusts the old boom box and presses play. We take the first breath in. Then the second.

The wave comes, washes over and runs up, then turns around and recedes, back to the ocean.

ACKNOWLEDGMENTS

o——o

The human body is a complicated subject. And how that body ingests, processes, and draws energy from air, and how that air affects our brains, bones, blood, bladders, and everything else . . . well, I learned over the last few years that understanding—and writing about—all that is another beast altogether.

I am deeply indebted to the medical pulmonauts who offered me their time, wisdom, coaching, and repeated respiratory rectifications along this wild and weird journey. Thank you, Dr. Jayakar Nayak, at the Stanford University Otolaryngology Head & Neck Surgery Center for slipping away from a ten-hour brain surgery to jam endoscopes up my nose and then explain the subtleties of cilia and sphenoids and sebaceous glands over salads at Vino Enoteca. (And a big thanks to Nayak's lab assistants, Nicole Borchard and Sachi Dholakia, for managing the mucosal madness.) Thank you, Dr. Marianna Evans, for schooling me in the ways of dysevolution and driving me all around Philly in such a nice car. Dr. Theodore Belfor and Dr. Scott Simonetti shared countless meals over countless months to describe the countless marvels of masticatory stress, nitric oxide, and Italian wines. Dr. Justin Feinstein at the Laureate Institute of Brain Research played hooky from his NIH lab work to give me a hard lesson in brain science, amygdalae, and the panicking power of carbon dioxide.

I begged and borrowed (with annotations, mind you) from several dozen elucidating books, interviews, and scientific articles penned by these respiratory renegades: Dr. Michael Gelb; Dr. Mark Burhenne; Dr. Steven Lin; Dr. Kevin Boyd; Dr. Ira Packman; Dr. John Feiner at the University of California, San Francisco Hypoxia Research Laboratory; Dr. Steven Park at the Department of Otorhinolaryngology, Albert Einstein College of Medicine; Dr. Amit Anand at the Division of Pulmonary, Critical Care, and Sleep Medicine at the Beth Israel Deaconess Medical Center; Ann Kearney, doctor of speech-language pathology at the Stanford Voice and Swallowing Center; and, of course, the generous and garrulous Drs. John and Mike Mew.

A crew of DIY pulmonauts welcomed me into their lives and lungs and showed me the transformative applications of breathing for real people living in the real world. Thank you to Chuck McGee III at Iced Viking Breathworks; Lynn Martin at MDH Breathing Coordination; Sasha Yakovleva at The Breathing Center; Luis Sérgio Álvares DeRose, John Conway Chisenhall, and Heduan Pinheiro at DeRose Method; Zach Fletcher at MindBodyClimb; and Tad Panther. *Un grand merci* to the mysterious and heretofore nameless clan of cataphiles for leading me deep below the Cimetière du Montparnasse and staining my jeans with thousand-year-old human bone dust. Thanks too to Mark Goettling at Bodimetrics for the suite of sleep and fitness monitoring devices and to Elizabeth Asch for offering up her luxurious Parisian pied-à-terre for an entire month.

Writing a paltry *tack så jävla mycket* feels like thin gratitude for my partner in nasal crime, Anders Olsson, a pulmonaut so dedicated to his craft that he ditched the glory of the Swedish *midsommar* to spend a month in dewy San Francisco with silicone up his nose, a pulse oximeter clipped to his finger, and tape on his lips. Thank you, Anders. But next time, can we just plug our ears instead?

My early attempts to leave no stone unturned in the lost art and science of breathing left me with a pile of word rubble. This is a long way of saying that this book, like most books, took a while and too often felt like Sisyphean drudgery.

Courtney Young, my virtuosic, adroit, and oft hilarious editor at Riverhead, boiled down a 270,000-word quagmire of sesquipedalian verbiage to the more digestible brick you now hold in your hands. Copilot/literary agent Danielle Svetcov at Levine Greenberg Rostan Literary Agency not only immediately returned my whiny calls—unheard-of in that line of work, trust me on this—but also worked side by side with me to sculpt, hone, and polish the verbs in her own ruthless and fabulous way. (Svetcov's constant support is priceless, or, at minimum, worth way more than that 15 percent.) Alex Heard, once again, winced while he whittled and whetted innumerable chapter drafts, down to the "dagger quotes," in almost legible cursive writing. (Sorry for ruining so many weekends, Alex.) Daniel Crewe at Penguin Books UK offered sage words of advice and encouragement early on and until the end.

I owe mafia-level favors to the readers who provided a much-needed editorial pummeling on early versions of this tome. Thank you to the crabby and scrupulous Adam Fisher; exclamatory Caroline Paul; poetic Matthew Zapruder; careful Michael Shryzpeck; adamant Richard Lowe; flexible Ron Penna; and cold-hearted Jason Dearen. Just give me a call whenever you need to move that body from the trunk.

My research assistant and fact-checker extraordinaire, Patrycja Przełucka, scoured through several hundred scientific papers with terrible titles like "The Correlation Between Erythropoiesis and Thrombopoiesis as an Index for Pre-Operative Autologous Blood Donation" and "Trained Breathing-Induced Oxygenation Acutely Reverses Cardiovascular Autonomic Dysfunction in Patients with Type 2 Diabetes and Renal Disease"—and then, *then*, she suffered the indignity of cross-checking this convolution, digit by digit, in the final drafts. Thank you, Patrycja, for your fastidiousness and spectacular grammar.

Lastly, *firstly*, to my lovely wife, Katie Storey, who breathes a constant supply of fresh, often-eucalyptus-scented air into my small office and frenzied life. *Vi ĉiam spiras freŝan aeron, varma hundo.*

Breath was written between the stacks of Weimar-era art books at the Mechanics' Institute Library in San Francisco, at the American Library in Paris, and on the kitchen table of that little red-doored house beside the old Catholic cemetery in Volcano, California, population 103.

BREATHING METHODS

o——o

Video and audio tutorials of these techniques, and more, are available at mrjamesnestor.com/breath.

CHAPTER 3. ALTERNATE NOSTRIL BREATHING (NADI SHODHANA)

This standard pranayama technique improves lung function and lowers heart rate, blood pressure, and sympathetic stress. It's an effective technique to employ before a meeting, an event, or sleep.

- (Optional) Hand Positioning: Place the thumb of your right hand gently over your right nostril and the ring

finger of that same hand on the left nostril. The forefinger
and middle finger should rest between the eyebrows.

- Close the right nostril with the thumb and inhale through
the left nostril very slowly.
- At the top of the breath, pause briefly, holding both nostrils closed, then lift just the thumb to exhale through the
right nostril.
- At the natural conclusion of the exhale, hold both nostrils
closed for a moment, then inhale through the right
nostril.
- Continue alternating breaths through the nostrils for five to
ten cycles.

CHAPTER 4. BREATHING COORDINATION

This technique helps to engage more movement from the diaphragm
and increase respiratory efficiency. It should never be forced; each breath
should feel soft and enriching.

- Sit up so that the spine is straight and chin is perpendicular to the body.
- Take a gentle breath in through the nose. At the top of the
breath begin counting softly aloud from one to 10 over
and over (*1, 2, 3, 4, 5, 6, 7, 8, 9, 10; 1, 2, 3, 4, 5, 6, 7, 8, 9, 10*).
- As you reach the natural conclusion of the exhale, keep
counting but do so in a whisper, letting the voice softly
trail out. Then keep going until only the lips are moving
and the lungs feel completely empty.
- Take in another large and soft breath and repeat.
- Continue for anywhere from 10 to 30 or more cycles.

Once you feel comfortable practicing this technique while sitting, try it while walking or jogging, or during other light exercise. For classes and individual coaching, visit http://www.breathingcoordination.ch /training.

CHAPTER 5. RESONANT (COHERENT) BREATHING

A calming practice that places the heart, lungs, and circulation into a state of coherence, where the systems of the body are working at peak efficiency. There is no more essential technique, and none more basic.

- Sit up straight, relax the shoulders and belly, and exhale.
- Inhale softly for 5.5 seconds, expanding the belly as air fills the bottom of the lungs.
- Without pausing, exhale softly for 5.5 seconds, bringing the belly in as the lungs empty. Each breath should feel like a circle.
- Repeat at least ten times, more if possible.

Several apps offer timers and visual guides. My favorites are *Paced Breathing* and *My Cardiac Coherence*, both of which are free. I try to practice this technique as often as possible.

BUTEYKO BREATHING

The point of Buteyko techniques is to train the body to breathe in line with its metabolic needs. For the vast majority of us, that means breathing less. Buteyko had an arsenal of methods, and almost all of them are based on extending the time between inhalations and exhalations, or breathholding. Here are a few of the simplest.

Control Pause

A diagnostic tool to gauge general respiratory health and breathing progress.

- Place a watch with a second hand or mobile phone with a stopwatch close by.
- Sit up with a straight back.
- Pinch both nostrils closed with the thumb and forefinger of either hand, then exhale softly out your mouth to the natural conclusion.
- Start the stopwatch and hold the breath.
- When you feel the first potent desire to breathe, note the time and take a soft inhale.

It's important that the first breath in after the Control Pause is controlled and relaxed; if it's labored or gasping, the breathhold was too long. Wait several minutes and try it again. The Control Pause should only be measured when you're relaxed and breathing normally, never after strenuous exercise or during stressed states. And like all breath restriction techniques, never attempt it while driving, while underwater, or in any other conditions where you might be injured should you become dizzy.

Mini Breathholds

A key component to Buteyko breathing is to practice breathing less all the time, which is what this technique trains the body to do. Thousands of Buteyko practitioners, and several medical researchers, swear by it to stave off asthma and anxiety attacks.

- Exhale gently and hold the breath for half the time of the Control Pause. (For instance, if the Control Pause is 40 seconds, the Mini Breathhold would be 20.)
- Repeat from 100 to 500 times a day.

Setting up timers throughout the day, every 15 minutes or so, can be helpful reminders.

Nose Songs

Nitric oxide is a powerhouse molecule that widens capillaries, increases oxygenation, and relaxes the smooth muscles. Humming increases the release of nitric oxide in the nasal passages 15-fold. This is the most effective, and simple, method for increasing this essential gas.

- Breathe normally through the nose and hum, any song or sound.
- Practice for at least five minutes a day, more if possible.

It may sound ridiculous, and feel ridiculous, and annoy those nearby, but the effects can be potent.

Walking/Running

Less extreme hypoventilation exercises (other than the misery I experienced jogging in Golden Gate Park) offer many of the benefits of high-altitude training. They are easy and can be practiced anywhere.

- Walk or run for a minute or so while breathing normally through the nose.
- Exhale and pinch the nose closed while keeping the same pace.
- When you sense a palpable air hunger, release the nose and breathe very gently, at about half of what feels normal for about 10 to 15 seconds.
- Return to regular breathing for 30 seconds.
- Repeat for about ten cycles.

Decongest the Nose

- Sit up straight and exhale a soft breath, then pinch both nostrils shut.
- Try to keep your mind off the breathholding; shake your head up and down or side to side; go for a quick walk, or jump and run.
- Once you feel a very potent sense of air hunger, take a very slow and controlled breath in through the nose. (If the nose is still congested, breathe softly through the mouth with pursed lips.)
- Continue this calm, controlled breathing for at least 30 seconds to 1 minute.
- Repeat all these steps six times.

Patrick McKeown's book *The Oxygen Advantage* offers detailed instructions and training programs in breathing less. Personalized instruction in Buteyko's method is available through www.consciousbreathing .com, www.breathingcenter.com, www.buteykoclinic.com, and with other certified Buteyko instructors.

CHAPTER 7. CHEWING

Hard chewing builds new bone in the face and opens airways. But for most of us, gnawing several hours a day—the amount of time and effort it takes to get such benefits—isn't possible, or preferable. A number of devices and proxies can fill the gap.

Gum

Any gum chewing can strengthen the jaw and stimulate stem cell growth, but harder textured varieties offer a more vigorous workout.

- Falim, a Turkish brand, is as tough as shoe leather and each piece lasts for about an hour. I've found the Sugarless Mint to be the most palatable. (Other flavors, such as Carbonate, Mint Grass, and sugar-filled varieties, tend to be softer and grosser.)
- Mastic gum, which comes from the resin of the evergreen shrub *Pistacia lentiscus*, has been cultivated in the Greek islands for thousands of years. Several brands are available through online retailers. The stuff can taste nasty but offers a rigorous jaw workout.

Oral Devices

As of this writing, Ted Belfor and his colleague, Scott Simonetti, received FDA clearance for a device called the POD (Preventive Oral Device), a small retainer that fits along the bottom row of teeth and simulates chewing stress. For more information, see www.discoverthepod.com and www.drtheodorebelfor.com.

Palatal Expansion

There are dozens of devices to expand the palate and open airways, each with its own advantages and disadvantages. Begin by contacting a dental professional who specializes in functional orthodontics.

Dr. Marianna Evans's Infinity Dental Specialists (at http://www .infinitydentalspecialists.com/) on the East Coast, and Dr. William Hang's Face Focused (https://facefocused.com) on the West Coast are among the most well-known and respected clinics in the United States, and good places to start. Across the pond, Britons can contact Dr. Mike Mew's clinic at https://orthodontichealth.co.uk.

CHAPTER 8. TUMMO

There are two forms of Tummo—one that stimulates the sympathetic nervous system, and another which triggers a parasympathetic response. Both work, but the former, made popular by Wim Hof, is much more accessible.

It's worth mentioning again that this technique should never be practiced near water, or while driving or walking, or in any other cir-cumstances where you might get hurt should you pass out. Consult your doctor if you are pregnant or have a heart condition.

- Find a quiet place and lie flat on your back with a pillow under the head. Relax the shoulders, chest, legs.
- Take 30 very deep, very fast breaths into the pit of the stomach and let it back out. If possible, breathe through the nose; if the nose feels obstructed, try pursed lips. The movement of each inhalation should look like a wave, fill-ing up in the stomach and softly moving up through the

lungs. Exhales follow the same movement, first emptying the stomach then the chest as air pours through the nose or pursed lips of the mouth.

- At the end of 30 breaths, exhale to the "natural conclusion," leaving about a quarter of the air in the lungs. Hold that breath for as long as possible.
- Once you've reached your absolute breathhold limit, take one huge inhale and hold it another 15 seconds. Very gently, move that fresh breath around the chest and to the shoulders, then exhale and start the heavy breathing again.
- Repeat the entire pattern at least three times.

Tummo takes some practice, and learning it from written instructions can be confusing and difficult. Chuck McGee, the Wim Hof Method instructor, offers free online sessions every Monday night at 9:00, Pacific Time. Sign up at https://www.meetup.com/Wim-Hof-Method -Bay-Area or log in through the Zoom platform: https://tinyurl.com /y4qwl3pm. McGee also offers personalized instruction throughout Northern California: https://www.wimhofmethod.com/instructors/chuck mcgee-iii.

Instructions for the calming version of Tummo meditation can be found at www.thewayofmeditation.com.au/revealing-the-secrets-of -tibetan-inner-fire-meditation.

CHAPTERS 9–10. SUDARSHAN KRIYA

This is the most powerful technique I've learned, and one of the most involved and difficult to get through. Sudarshan Kriya consists of four phases: *Om* chants, breath restriction, paced breathing (inhaling for 4 seconds, holding for 4 seconds, exhaling for 6, then holding for 2), and, finally, 40 minutes of very heavy breathing.

A few YouTube tutorials are available, but to get the motions correct, deeper instruction is highly recommended. The Art of Living offers weekend workshops to guide new students through the practice. See more at www.artofliving.org.

. . .

Below are several breathing practices that didn't make the cut in the main text of this book for one reason or another. I regularly practice them, as do millions of others. Each is useful and powerful in its own way.

Yogic Breathing (Three-Part)

A standard technique for any aspiring pranayama student.

PHASE I
- Sit in a chair or cross-legged and upright on the floor and relax the shoulders.
- Place one hand over the navel and slowly breathe into the belly. You should feel the belly expand with each breath in, deflate with each breath out. Practice this a few times.
- Next, move the hand up a few inches so that it's covering the bottom of the rib cage. Focus the breath into the location of the hand, expanding the ribs with each inhale, retracting them with each exhale. Practice this for about three to five breaths.
- Move the hand to just below the collarbone. Breathe deeply into this area and imagine the chest spreading out and withdrawing with each exhale. Do this for a few breaths.

PHASE II

- Connect all these motions into one breath, inhaling into the stomach, lower rib cage, then chest.
- Exhale in the opposite direction, first emptying the chest, then the rib cage, then the stomach. Feel free to use a hand and feel each area as you breathe in and out of it.
- Continue this same sequence for about a dozen rounds.

These motions will feel very awkward at first, but after a few breaths they get easier.

Box Breathing

Navy SEALs use this technique to stay calm and focused in tense situations. It's simple.

- Inhale to a count of 4; hold 4; exhale 4; hold 4. Repeat.

Longer exhalations will elicit a stronger parasympathetic response. A variation of Box Breathing to more deeply relax the body that's especially effective before sleeping is as follows:

- Inhale to a count of 4; hold 4; exhale 6; hold 2. Repeat.

Try at least six rounds, more if necessary.

Breathhold Walking

Anders Olsson uses this technique to increase carbon dioxide and, thus, increase circulation in his body. It's not much fun, but the benefits, Olsson told me, are many.

- Go to a grassy park, beach, or anywhere else where the ground is soft.
- Exhale all the breath, then walk slowly, counting each step.
- Once you feel a powerful sense of air hunger, stop counting and take a few very calm breaths through the nose while still walking. Breathe normally for at least a minute, then repeat the sequence.

The more you practice this technique, the higher the count. Olsson's record is 130 steps; mine is about a third of that.

4-7-8 Breathing

This technique, made famous by Dr. Andrew Weil, places the body into a state of deep relaxation. I use it on long flights to help fall asleep.

- Take a breath in, then exhale through your mouth with a *whoosh* sound.
- Close the mouth and inhale quietly through your nose to a mental count of four.
- Hold for a count of seven.
- Exhale completely through your mouth, with a *whoosh*, to the count of eight.
- Repeat this cycle for at least four breaths.

Weil offers a step-by-step instructional on YouTube, which has been viewed more than four million times. https://www.youtube.com/watch?v=gz4G31LGyog.

NOTES

For a full bibliography with updated and expanded notes, please visit
mrjamesnestor.com/breath.

ix **"In transporting the breath":** *Primordial Breath: An Ancient Chinese Way of Prolonging Life through Breath Control*, vol. 1, *Seven Treatises from the Taoist Canon, the Tao Tsang, on the Esoteric Practice of Embryonic Breathing*, trans. Jane Huang and Michael Wurmbrand, 1st ed. (Original Books, 1987), 3.

Introduction

xvi **master the art of breathing:** I wrote about freediving and the human connection to the sea in my first book, *Deep* (New York: Houghton Mifflin Harcourt, 2014).

xvii **books of the Chinese Tao:** *The Primordial Breath: An Ancient Chinese Way of Prolonging Life through Breath Control*, vol. 1, *Seven Treatises from the Taoist Canon, the Tao Tsang, on the Esoteric Practice of Embryonic Breathing*, trans. Jane Huang and Michael Wurmbrand, 1st ed. (Original Books, 1987); Christophe André, "Proper Breathing Brings Better Health," *Scientific American*, Jan. 15, 2019; Bryan Gandevia, "The Breath of Life: An Essay on the Earliest History of Respiration: Part II," *Australian Journal of Physiotherapy* 16, no. 2 (June 1970): 57–69.

xviii **ancient Tao text:** *The Primordial Breath*, 8.

xviii **confirmed this position:** In the December 1998 issue of *The New Republic*, the editor of the *New England Journal of Medicine* argued that health determines how we breathe, and how we breathe has no effect on the state of health. In the introduction to Teresa Hale's book *Breathing Free: The Revolutionary 5-Day Program to Heal Asthma, Emphysema, Bronchitis, and Other Respiratory Ailments* (New York: Harmony, 1999), Dr. Leo Galland, Fellow of the American College of Nutrition and the American College of Physicians, described exactly how the ways in which we breathe directly affect health. Galland's account was one of several that I discovered in the initial research for this book and subsequent conversations with professors, doctors, and others in the medical field.

Chapter One: The Worst Breathers in the Animal Kingdom

3 **dental arches and sinus cavity:** Karina Camillo Carrascoza et al., "Consequences of Bottle-Feeding to the Oral Facial Development of Initially Breastfed Children," *Jornal de Pediatria* 82, no. 5 (Sept.–Oct. 2006): 395–97.

4 **increasing his chances of developing:** A retrospective review of more than 7,300 adults associated a 2 percent higher risk of obstructive sleep apnea with every tooth lost. If five to eight teeth were removed, that percentage increased to 25 percent; nine to 31 teeth showed a 36 percent increase. Those patients who had all their teeth removed suffered a 60 percent greater chance of acquiring sleep apnea. Anne E. Sanders et al., "Tooth Loss and Obstructive Sleep Apnea Signs and Symptoms in the US Population," *Sleep Breath* 20, no. 3 (Sept. 2016): 1095–102. Related studies: Derya Germeç-Çakan et al., "Uvulo-Glossopharyngeal Dimensions in Non-Extraction, Extraction with Minimum Anchorage, and Extraction with Maximum Anchorage," *European Journal of Orthodontics* 33, no. 5 (Oct. 2011): 515–20; Yu Chen et al., "Effect of Large Incisor Retraction on Upper Airway Morphology in Adult Bimaxillary Protrusion Patients: Three-Dimensional Multislice Computed Tomography Registration Evaluation," *The Angle Orthodontist* 82, no. 6 (Nov. 2012): 964–70.

4 **Twenty-five sextillion molecules:** Simon Worrall, "The Air You Breathe Is Full of Surprises," *National Geographic*, Aug. 13, 2012, https://www.nationalgeographic.com/news/2017/08/air-gas-caesar-last-breath-sam-kean.

5 **around half of us:** Mouthbreathing estimates are murky and range from 5 to 75 percent. Two independent studies in Brazil showed that more than 50 percent of children are mouthbreathers, but the condition may be more common than that. Valdenice Aparecida de Menezes et al., "Prevalence and Factors Related to Mouth Breathing in School Children at the Santo Amaro Project—Recife, 2005," *Brazilian Journal of Otorhinolaryngology* 72, no. 3 (May–June 2006): 394–98; Rubens Rafael Abreu et al., "Prevalence of Mouth Breathing among Children," *Jornal de Pediatria* 84, no. 5 (Sept.–Oct. 2008): 467–70; Michael Stewart et al., "Epidemiology and Burden of Nasal Congestion," *International Journal of General Medicine* 3 (2010): 37–45; David W. Hsu and Jeffrey D. Suh, "Anatomy and Physiology of Nasal Obstruction," *Otolaryngologic Clinics of North America* 51, no. 5 (Oct. 2018): 853–65.

6 **The causes are many:** "Symptoms: Nasal Congestion," Mayo Clinic, https://www.mayoclinic.org/symptoms/nasal-congestion/basics/causes/sym-20050644.

6 **When mouths don't grow:** Michael Friedman, ed., *Sleep Apnea and Snoring: Surgical and Non-Surgical Therapy*, 1st ed. (Philadelphia: Saunders/Elsevier, 2009), 6.

9 **4 billion years ago:** Keith Cooper, "Looking for LUCA, the Last Universal Common Ancestor," Astrobiology at NASA: Life in the Universe, Mar. 17, 2017, https://astrobiology.nasa.gov/news/looking-for-luca-the-last-universal-common-ancestor/.

9 **oxygen waste in the atmosphere:** "New Evidence for the Oldest Oxygen-Breathing Life on Land," ScienceDaily, Oct. 21, 2011, https://www.sciencedaily.com/releases/2011/10/111019181210.htm.

9 **16 times more energy:** S. E. Gould, "The Origin of Breathing: How Bacteria Learnt to Use Oxygen," *Scientific American*, July 29, 2012, https://blogs.scientificamerican.com/lab-rat/the-origin-of-breathing-how-bacteria-learnt-to-use-oxygen.

12 **straight teeth:** Not all the skulls had teeth. But Evans and Boyd could tell from the shape of the jaw and tooth cavities that the teeth that had been there were straight.

13 **Harvard biologist Daniel Lieberman:** Lieberman defines dysevolution as "the deleterious feedback loop that occurs over multiple generations when we don't treat the causes of a mismatch disease but instead pass on whatever environmental factors cause the disease, keeping the disease prevalent and sometimes making it worse." A "mismatch disease" begins "when we get sick or injured from an evolutionary mismatch that results from being inadequately adapted to a change in the body's environment." You can read more about dysevolution in Lieberman's book *The Story of the Human Body: Evolution, Health, and Disease* (New York: Pantheon, 2013); the quote is from p. 176. See also Jeff Wheelwright, "From Diabetes to Athlete's Foot, Our Bodies Are Maladapted for Modern Life," *Discover*, Apr. 2, 2015, http://discovermagazine.com/2015/may/16-days-of-dysevolution.

13 **sharp enough to carve tongues:** Briana Pobiner, "The First Butchers," *Sapiens*, Feb. 23, 2016, https://www.sapiens.org/evolution/homo-sapiens-and-tool-making.

13 **Tenderizing food:** Daniel E. Lieberman, *The Evolution of the Human Head* (Cambridge, MA: Belknap Press of Harvard University Press, 2011), 255–81.

13 **Grilling food:** For instance, animals can use only 50 to 60 percent of the nutrients from a raw egg but more than 90 percent from a cooked egg. The same is true with many cooked plants, vegetables, and meats. Steven Lin, *The Dental Diet: The Surprising Link between Your Teeth, Real Food, and Life-Changing Natural Health* (Carlsbad, CA: Hay House, 2018), 35.

13 **800,000 years ago:** Likely much earlier. At the Koobi Fora in Kenya, researchers found evidence of a fire that had been intentionally made 1.6 million years ago. Amber Dance, "Quest for Clues to Humanity's First Fires," *Scientific American*, June 19, 2017, https://www.scientificamerican.com/article/quest-for-clues-to-humanitys-first-fires; Kenneth Miller, "Archaeologists Find Earliest Evidence of Humans Cooking with Fire," *Discover*, Dec. 17, 2013, http://discovermagazine.com/2013/may/09-archaeologists-find-earliest-evidence-of-humans-cooking-with-fire.

14 **saved even more energy:** How much brain did we gain from a smaller gut? Nobody really knows for sure, but it's significant. An exhaustive overview is available in Leslie C. Aiello, "Brains and Guts in Human Evolution: The Expensive Tissue Hypothesis," Mar. 1997, http://www.scielo.br/scielo.php?script=sci_arttext&pid=S0100-84551997000100023.

14 **50 percent larger:** Richard Wrangham, a Harvard University biological anthropologist, has studied ancient hominin diets extensively. Read more from various perspectives: Rachel Moeller, "Cooking Up Bigger Brains," *Scientific American*, Jan. 1, 2008, https://www.scientificamerican.com/article/cooking-up-bigger-brains.

14 **second glance:** "Did Cooking Give Humans an Evolutionary Edge?," NPR, Aug. 28, 2009, https://www.npr.org/templates/story/story.php?storyId=112334465.

14 **vertically positioned nose:** Colin Barras, "The Evolution of the Nose: Why Is the Human Hooter So Big?," *New Scientist*, Mar. 24, 2016, https://www.newscientist.com/article/2082274-the-evolution-of-the-nose-why-is-the-human-hooter-so-big/; "Mosaic Evolution of Anatomical Foundations of Speech," Systematics & Phylogeny Section, Primate Research Institute, Kyoto University. Nishimura Lab, https://www.pri.kyoto-u.ac.jp/shinka/keitou/nishimura-HP/tn_res-e.html.

14 **the tighter our airways became:** "The surface area of their nasal cavity is about half of what the scaling suggests, and the volume is even only about 10% of the prediction. . . . In fact, the volume of the human nasal cavity is almost 90% smaller than expected." David Zwickler, "Physical and Geometric Constraints Shape the Labyrinth-like Nasal Cavity," *Proceedings of the National Academy of Sciences*, Jan. 26, 2018.

14 **make clothes:** Colin Barras, "Ice Age Fashion Showdown: Neanderthal Capes Versus Human Hoodies," *New Scientist*, Aug. 8, 2016, https://www.newscientist.com/article/2100322-ice-age-fashion-showdown-neanderthal-capes-versus-human-hoodies/.

14 *Homo naledi:* "Homo Naledi," Smithsonian National Museum of Natural History,

http://humanorigins.si.edu/evidence/human-fossils/species/homo-naledi.

15 **we adapted wider and flatter noses:** Ben Panko, "How Climate Helped Shape Your Nose," Smithsonian.com, Mar. 16, 2017, https://www.smithsonianmag.com/science-nature/how-climate-changed-shape-your-nose-180962567.

15 **more efficient at inhaling:** Joan Raymond, "The Shape of a Nose," *Scientific American*, Sept. 1, 2011, https://www.scientificamerican.com/article/the-shape-of-a-nose.

15 **larynx sank:** Whether enabling speech was the driving factor or a lucky by-product, for one reason or another the *Homo sapiens* larynx sank. Asif A. Ghazanfar and Drew Rendall, "Evolution of Human Vocal Production," *Current Biology* 18, no. 11 (2008): R457–60, https://www.cell.com/current-biology/pdf/S0960-9822(08)00371-0.pdf; Kathleen Masterson, "From Grunting to Gabbing: Why Humans Can Talk," NPR, Aug. 11, 2010, https://www.npr.org/templates/story/story.php?storyId=129083762.

15 **wider range of vocalizations:** How much this lowered larynx benefited early humans in developing a complex spoken language is hotly debated. Nobody knows for sure, but, as I've found, anthropologists are more than willing to offer up opinions. Ghazanfar and Rendall, "Evolution"; Lieberman, *Story of the Human Body*, 171–72.

16 **human species, that could easily choke:** Choking on food is the fourth leading cause of accidental deaths in the U.S. "We have paid a heavy price for speaking more clearly," wrote Daniel Lieberman, in *Story of the Human Body*, 144.

17 **Nasal obstruction triggers:** Terry Young et al., the University of Wisconsin Sleep and Respiratory Research Group, "Nasal Obstruction as a Risk Factor for Sleep-Disordered Breathing," *Journal of Allergy and Clinical Immunology* 99, no. 2 (Feb. 1997): S757–62; Mahmoud I. Awad and Ashutosh Kacker, "Nasal Obstruction Considerations in Sleep Apnea," *Otolaryngologic Clinics of North America* 51, no. 5 (Oct. 2018): 1003–1009.

Chapter Two: Mouthbreathing

21 **into a state of stress:** This blog entry includes a thorough explanation with 43

scientific references: "The Nose Knows: A Case for Nasal Breathing During High Intensity Exercise," Adam Cap website, https://adamcap.com/2013/11/29/the-nose-knows/.

24 **swore off breathing through their mouths:** More explanation from Douillard on the importance of nasal breathing in exercise: "Ayurvedic Fitness," John Douillard, PTonthenet, Jan. 3, 2007, https://www.ptonthenet.com/articles/Ayurvedic-Fitness-2783.

24 **body doesn't have enough oxygen:** A good, simple explanation of anaerobic and aerobic energies: Andrea Boldt, "What Is the Difference Between Lactic Acid & Lactate?," https://www.livestrong.com/article/470283-what-is-the-difference-between-lactic-acid-lactate/.

24 **an excess of lactic acid:** Stephen M. Roth, "Why Does Lactic Acid Build Up in Muscles? And Why Does It Cause Soreness?," *Scientific American*, Jan. 23, 2006, https://www.scientificamerican.com/article/why-does-lactic-acid-buil/.

25 **feeling of anaerobic overload:** Anaerobic exhaustion, and its associated lactic acidosis, isn't always triggered by strenuous exercise. It can also occur through liver disease, alcoholism, severe trauma, or other conditions that deprive the body of the oxygen it needs to function aerobically. Lana Barhum, "What to Know About Lactic Acidosis," *Medical News Today*, https://www.medicalnewstoday.com/articles/320863.php.

25 **anaerobic muscle fibers:** Human muscle fibers are an interwoven mixture of aerobic and anaerobic fibers, whereas other animals, such as chickens, have entire muscle systems that are either aerobic or anaerobic. The dark meat in a cooked chicken is dark because these muscles were used to provide aerobic energy and are filled with oxygenated blood; white meat is anaerobic, and so is lacking in these red pigments. Phillip Maffetone, *The Maffetone Method: The Holistic, Low-Stress, No-Pain Way to Exceptional Fitness* (Camden, ME: Ragged Mountain Press/McGraw-Hill, 1999), 21.

25 **eventually break down:** Dr. Valter Longo, director of the Longevity Institute at the University of Southern California–Davis

School of Gerontology, offers some interesting perspective here: https://www
.bluezones.com/2018/01/what-exercise
-best-happy-healthy-life/.

25 **37 trillion:** Eva Bianconi et al., "An Estimation of the Number of Cells in the Human Body," *Annals of Human Biology* 40, no. 6 (Nov. 2013): 463–71.

25 **16 times more energy efficiency:** The actual numbers work out to 2 ATPs per glucose molecule for anaerobic energy and 38 ATPs per glucose molecule for aerobic energy. For this reason, most textbooks say that aerobic energy has a 19-fold increase over anaerobic energy. But what most textbooks don't account for is inefficiencies and waste in the ATP process, which usually suck up about 8 ATPs. A more conservative estimate, then, is that aerobic respiration produces something closer to 30 to 32 ATPs, or around 16 times the energy that anaerobic produces. Peter R. Rich, "The Molecular Machinery of Keilin's Respiratory Chain," *Biochemical Society Transactions* 31, no. 6 (Dec. 2003): 1095–105.

25 **standardized workouts could be more injurious:** To be clear, Maffetone never argued against occasionally entering into anaerobic exercise. Rowing, lifting weights, and running can all have a profound effect on strength and endurance. But to be effective, these exercises needed to be kept in context of larger training, and can't be prioritized over aerobic training. High-intensity interval training only works because well-designed programs are built around spending the vast majority of time in periods of slower, gentler aerobic exercise. Author and fitness trainer Brian MacKenzie argues that the key to high levels of performance fitness is combining aerobic and anaerobic exercise effectively. *The Maffetone Method*, 56; Brian MacKenzie with Glen Cordoza, *Power Speed Endurance: A Skill-Based Approach to Endurance Training* (Las Vegas: Victory Belt, 2012), Kindle locations 462–70; Alexandra Patillo, "You're Probably Doing Cardio All Wrong: 2 Experts Reveal How to Train Smarter," *Inverse*, Aug. 7, 2019, https://www.inverse.com/article/58370
-truth-about-cardio?refresh=39.

26 **subtract your age from 180:** Those with heart disease or other medical conditions should subtract 10 from Maffetone's equation; if you have asthma or allergies or have not exercised before, subtract 5. Competitive athletes who have been training for more than two years, add 5. This works out to around 80 percent of maximum capacity for a man of my age. Anaerobic states usually hit at 80 percent, or the stage at which it becomes difficult to speak in full sentences. "Know Your Target Heart Rates for Exercise, Losing Weight and Health," Heart.org, https://www.heart.org/en/healthy-living/fitness
/fitness-basics/target-heart-rates; Wendy Bumgardner, "How to Reach the Anaerobic Zone during Exercise," VeryWellFit, Aug. 30, 2019, https://www.verywellfit
.com/anaerobic-zone-3436576.

26 **below this rate but never above it:** Two thousand years ago, a Chinese surgeon named Hua Tuo prescribed only moderate exercise to his patients, warning them: "The body needs exercise, only it must not be to the point of exhaustion, for exercise expels the bad air in the system, promotes free circulation of the blood, and prevents sickness." The most efficient state of exercise where we reap the most benefits, Maffetone found, was around or below 60 percent of maximum capacity. The Cooper Institute, a research foundation that for 50 years has been studying the links between physical activity and chronic disease, has found that exercising at 50 percent leads to massive gains in aerobic fitness, improved blood pressure, prevention of various diseases, and more. Several other studies over the past several decades confirm this. Meanwhile, overexercising above 60 percent, toward that anaerobic zone, has been shown to induce a stress state, increased cortisol, adrenaline, and oxidative stress. Charles M. Tipton, "The History of 'Exercise Is Medicine' in Ancient Civilizations," *Advances in Physiology Education*, June 2014, 109–17; Helen Thompson, "Walk, Don't Run," *Texas Monthly*, June 1995, https://www
.texasmonthly.com/articles/walk
-dont-run; Douillard, *Body, Mind, and Sport*, 205; Chris E. Cooper et al., "Exercise, Free Radicals and Oxidative Stress," *Biochemical Society Transactions* 30, part 2 (May 2002): 280–85.

27 **troop of rhesus monkeys:** Peter A. Shapiro, "Effects of Nasal Obstruction on Facial Development," *Journal of Allergy and Clinical Immunology* 81, no. 5, part 2 (May 1988): 968; Egil P. Harvold et al., "Primate Experiments on Oral Sensation and Dental Malocclusions," *American Journal of Orthodontics & Dentofacial Orthopedics* 63, no. 5 (May 1973): 494–508; Egil P. Harvold et al., "Primate Experiments on Oral Respiration," *American Journal of Orthodontics* 79, no. 4 (Apr. 1981): 359–72; Britta S. Tomer and E. P. Harvold, "Primate Experiments on Mandibular Growth Direction," *American Journal of Orthodontics* 82, no. 2 (Aug. 1982): 114–19; Michael L. Gelb, "Airway Centric TMJ Philosophy," *Journal of the California Dental Association* 42, no. 8 (Aug. 2014): 551–62; Karin Vargervik et al., "Morphologic Response to Changes in Neuromuscular Patterns Experimentally Induced by Altered Modes of Respiration," *American Journal of Orthodontics* 85, no. 2 (Feb. 1984): 115–24.

27 **happens to our own species:** Yu-Shu Huang and Christian Guilleminault, "Pediatric Obstructive Sleep Apnea and the Critical Role of Oral-Facial Growth: Evidences," *Frontiers in Neurology* 3, no. 184 (2012), https://www.frontiersin.org/articles/10.3389/fneur.2012.00184/full; Anderson Capistrano et al., "Facial Morphology and Obstructive Sleep Apnea," *Dental Press Journal of Orthodontics* 20, no. 6 (Nov.–Dec. 2015): 60–67.

27 **changes the physical body:** A few of the better studies: Cristina Grippaudo et al., "Association between Oral Habits, Mouth Breathing and Malocclusion," *Acta Otorhinolaryngologica Italica* 36, no. 5 (Oct. 2016): 386–94; Yosh Jefferson, "Mouth Breathing: Adverse Effects on Facial Growth, Health, Academics, and Behavior," *General Dentistry* 58, no. 1 (Jan.–Feb. 2010): 18–25; Doron Harari et al., "The Effect of Mouth Breathing versus Nasal Breathing on Dentofacial and Craniofacial Development in Orthodontic Patients," *Laryngoscope* 120, no. 10 (Oct. 2010): 2089–93; Valdenice Aparecida de Menezes, "Prevalence and Factors Related to Mouth Breathing in School Children at the Santo Amaro Project—Recife, 2005," *Brazilian Journal of Otorhinolaryngology* 72, no. 3 (May–June 2006): 394–98.

27 **Patrick McKeown:** Patrick McKeown and Martha Macaluso, "Mouth Breathing: Physical, Mental and Emotional Consequences," Central Jersey Dental Sleep Medicine, Mar. 9, 2017, https://sleep-apnea-dentist-nj.info/mouth-breathing-physical-mental-and-emotional-consequences/.

28 **When seasonal allergies hit:** W. T. McNicholas, "The Nose and OSA: Variable Nasal Obstruction May Be More Important in Pathophysiology Than Fixed Obstruction," *European Respiratory Journal* 32 (2008): 5, https://erj.ersjournals.com/content/32/1/3; C. R. Canova et al., "Increased Prevalence of Perennial Allergic Rhinitis in Patients with Obstructive Sleep Apnea," *Respiration* 71 (Mar.–Apr. 2004): 138–43; Carlos Torre and Christian Guilleminault, "Establishment of Nasal Breathing Should Be the Ultimate Goal to Secure Adequate Craniofacial and Airway Development in Children," *Jornal de Pediatria* 94, no. 2 (Mar.–Apr. 2018): 101–3.

29 **obstructive sleep apnea:** Sleep apnea and snoring are common bedfellows. The more and louder we snore, the more the airways become damaged and the more susceptible we are to sleep apnea. Farhan Shah et al., "Desmin and Dystrophin Abnormalities in Upper Airway Muscles of Snorers and Patients with Sleep Apnea," *Respiratory Research* 20, no. 1 (Dec. 2019): 31.

29 **"More wholesome to sleep":** Levinus Lemnius, *The Secret Miracles of Nature: In Four Books* (London, 1658), 132–33, https://archive.org/details/b30326084/page/n7; Melissa Grafe, "Secret Miracles of Nature," Yale University, Harvey Cushing/John Hay Whitney Medical Library, Dec. 12, 2013, https://library.medicine.yale.edu/content/secret-miracles-nature.

29 **lose 40 percent more water:** Sophie Svensson et al., "Increased Net Water Loss by Oral Compared to Nasal Expiration in Healthy Subjects," *Rhinology* 44, no. 1 (Mar. 2006): 74–77.

29 **During the deepest, most restful:** Mark Burhenne, *The 8-Hour Sleep Paradox: How We Are Sleeping Our Way to Fatigue,*

Disease and Unhappiness (Sunnyvale, CA: Ask the Dentist, 2015), 45.

30 **vasopressin, which communicates:** Andrew Bennett Hellman, "Why the Body Isn't Thirsty at Night," *Nature News,* Feb. 28, 2010, https://www.nature.com/news /2010/100228/full/news.2010.95.html.

30 **report from the Mayo Clinic:** In 2001, researchers at the University of Pittsburgh surveyed several hundred people and found that half of those with insomnia also suffer from obstructive sleep apnea. Then they surveyed people with obstructive sleep apnea and found that half had insomnia. Years later, a study published in the *Mayo Clinic Proceedings* of 1,200 chronic insomniacs found that all 900 of the patients prescribed some kind of drug to help them sleep, including antidepressants, had "pharmocotherapeutic failure." The more than 700 patients taking prescription drugs reported the most severe insomnia. These drugs not only are ineffective for the patients taking them, but can actually make sleep quality worse because insomnia for many people isn't a psychological problem; it's a breathing problem. Barry Krakow et al., "Pharmacotherapeutic Failure in a Large Cohort of Patients with Insomnia Presenting to a Sleep Medicine Center and Laboratory: Subjective Pretest Predictions and Objective Diagnoses," *Mayo Clinic Proceedings* 89, no. 12 (Dec. 2014): 1608–20; "Pharmacotherapy Failure in Chronic Insomnia Patients," *Mayo Clinic Proceedings,* YouTube, https://youtube.com/watch ?v=vdm1kTFJCK4.

30 **millions of Americans:** Thomas M. Heffron, "Insomnia Awareness Day Facts and Stats," Sleep Education, Mar. 10, 2014, http://sleepeducation.org/news/2014/03 /10/insomnia-awareness-day-facts-and -stats.

30 **"increased respiratory effort":** Guillemainault argued that paying too-close attention to specific scores muddles the larger problem of snoring and sleep apnea. Any disturbances in breathing during sleep, be it apnea, snoring, heavy breathing, or even the slightest constriction in the throat, can reap heavy damage to the body. Christian Guilleminault and Ji Hyun Lee, "Does Benign 'Primary Snoring'

Ever Exist in Children?," *Chest Journal* 126, no. 5 (Nov. 2004): 1396–98; Guilleminault et al., "Pediatric Obstructive Sleep Apnea Syndrome," *Archives of Pediatrics and Adolescent Medicine* 159, no. 8 (Aug. 2005): 775–85.

30 **making me dumber:** Noriko Tsubamoto-Sano et al., "Influences of Mouth Breathing on Memory and Learning Ability in Growing Rats," *Journal of Oral Science* 61, no. 1 (2019): 119–24; Masahiro Sano et al., "Increased Oxygen Load in the Prefrontal Cortex from Mouth Breathing: A Vector-Based Near-Infrared Spectroscopy Study," *Neuroreport* 24, no. 17 (Dec. 2013): 935–40; Malia Wollan, "How to Be a Nose Breather," *The New York Times Magazine,* Apr. 23, 2019.

31 **The breath inhaled:** *The Primordial Breath: An Ancient Chinese Way of Prolonging Life through Breath Control,* vol. 2, trans. Jane Huang and Michael Wurmbrand (Original Books, 1990), 31.

32 **And here we are:** The stats for malocclusion vary. Kevin Boyd, a pediatric dentist, and Darius Loghmanee, a physician and sleep specialist, noted that "75% of children, ages 6 to 11 and 89% of youths, ages 12 to 17, have some degree of malocclusion." In addition, an estimated 65 percent of adults have some degree of malocclusion; this population includes those adults who have already had orthodontic procedures. Given this, the actual number of these adults had not received treatment would be closer to 90 percent. Other estimates I found put the figure for children even higher. Suffice to say, it's a lot. A few slide presentations (with references) and in-depth interviews about malocclusion: Kevin L. Boyd and Darius Loghmanee, "Inattention, Hyperactivity, Snoring and Restless Sleep: My Child's Dentist Can Help?!," presentation at 3rd Annual Autism, Behavior, and Complex Medical Needs Conference; Kevin Boyd interview by Shirley Gutkowski, Cross Link Radio, 2017, https://crosslinkradio.com/dr-kevin -boyd-2/; "Malocclusion," Boston Children's Hospital, http://www.childrenshos pital.org/conditions-and-treatments/con ditions/m/malocclusion.

32 **Forty-five percent of adults snore:** "Snoring," Columbia University Department of

Neurology, http://www.columbianeurol
ogy.org/neurology/staywell/document
.php?id=42066.

32 Twenty-five percent of American adults:
"Rising Prevalence of Sleep Apnea in U.S.
Threatens Public Health," press release,
American Academy of Sleep Medicine,
Sept. 29, 2014.

32 an estimated 80 percent: Steven Y. Park,
MD, *Sleep, Interrupted: A Physician Re-
veals the #1 Reason Why So Many of Us
Are Sick and Tired* (New York: Jodev
Press, 2008), 26.

32 than there were 10,000 years ago: Index
of world population estimates throughout the decades: https://tinyurl.com
/rrhvcjh.

33 slack-jawed and narrowed faces: Several
studies have shown similar restoration in
humans. In the 1990s, Canadian researchers measured the facial and mouth dimensions of 38 children suffering from
chronically enlarged adenoids, the glands
located on the roof of the mouth that help
fight infections. The swollen glands made
it almost impossible for the children to
breathe through their noses, so they all
adopted mouthbreathing, and all had the
long, slack-jawed, narrow-faced profiles
that come with it. Surgeons removed the
adenoids from half the children and mon-
itored the measurements of their faces.
Slowly, surely, their faces morphed back
into their natural position: jaws moved
forward, the maxilla flared outward. Donald C. Woodside et al., "Mandibular and
Maxillary Growth after Changed Mode of
Breathing," *American Journal of Orthodontics and Dentofacial Orthopedics* 100,
no. 1 (July 1991): 1–18; Shapiro, "Effects of
Nasal Obstruction on Facial Development," 967–68.

Chapter Three: Nose

38 Smell is life's oldest sense: Interview with
Dolores Malaspina, MD, professor of clinical psychiatry at Columbia University in
New York; Nancie George, "10 Incredible Facts about Your Sense of Smell,"
EveryDay Health, https://www.everyday
health.com/news/incredible-facts
-about-your-sense-smell/.

39 stores memories: Artin Arshamian et al.,
"Respiration Modulates Olfactory Memory

Consolidation in Humans," *Journal of
Neuroscience* 38, no. 48 (Nov. 2018):
10286–94; Christina Zelano et al., "Nasal
Respiration Entrains Human Limbic Os-
cillations and Modulates Cognitive Function," *Journal of Neuroscience* 36, no. 49
(Dec. 2016): 12448–67.

39 suffer from asthma: A. B. Ozturk et al.,
"Does Nasal Hair (Vibrissae) Density
Affect the Risk of Developing Asthma in
Patients with Seasonal Rhinitis?," *Inter-
national Archives of Allergy and Immunol-
ogy* 156, no. 1 (Mar. 2011): 75–80.

40 an Indian surgeon: Ananda Balayogi
Bhavanani, "A Study of the Pattern of Nasal Dominance with Reference to Different Phases of the Lunar Cycle," *Yoga Life*
35 (June 2004): 19–24.

40 called nasal cycles: Sometimes referred to
as an "ultradian rhythm," meaning a cycle
shorter than the period of circadian
rhythm.

40 first described in 1895: A comprehensive
review of the nasal cycle can be found in
Alfonso Luca Pendolino et al., "The Nasal
Cycle: A Comprehensive Review," *Rhinology Online* 1 (June 2018): 67–76; R. Kayser,
"Die exacte Messung der Luftdurchgängig-
keit der Nase," *Archives of Laryngology* 3
(1895): 101–20.

40 30 minutes to 4 hours: This is an esti-
mate. Some studies have shown that the
nasal cycle fluctuates between 30 minutes
and two and a half hours; others show the
cycle can last up to four hours. Roni
Kahana-Zweig et al., "Measuring and
Characterizing the Human Nasal Cycle,"
PloS One 11, no. 10 (Oct. 2016): e0162918;
Rauf Tahamiler et al., "Detection of the
Nasal Cycle in Daily Activity by Remote
Evaluation of Nasal Sound," *Archives of
Otolaryngology–Head and Neck Surgery*
129, no. 9 (Feb. 2009): 137–42.

41 "honeymoon rhinitis": "Sneezing 'Can
Be Sign of Arousal,'" BBC News, Dec. 19,
2008, http://news.bbc.co.uk/2/hi/health
/7792102.stm; Andrea Mazzatenta et al.,
"Swelling of Erectile Nasal Tissue Induced
by Human Sexual Pheromone," *Advances
in Experimental Medicine and Biology* 885
(2016): 25–30.

41 nostrils cycled: Kahana-Zweig et al.,
"Measuring"; Marc Oliver Scheithauer,
"Surgery of the Turbinates and 'Empty

Nose' Syndrome," *GMS Current Topics in Otorhinolaryngology–Head and Neck Surgery* 9 (2010): Doc3.

41 **body to flip over:** In addition, nasal cycles appear to be associated with the duration of deep sleep. A. T. Atanasov and P. D. Dimov, "Nasal and Sleep Cycle—Possible Synchronization during Night Sleep," *Medical Hypotheses* 61, no. 2 (Aug. 2003): 275–77; Akihira Kimura et al., "Phase of Nasal Cycle During Sleep Tends to Be Associated with Sleep Stage," *The Laryngoscope* 123, no. 6 (Aug. 2013): 1050–55.

41 **become inflamed:** Pendolino et al., "The Nasal Cycle."

41 **back and forth quickly:** A lagging nasal cycle in some cultures was considered a harbinger of disease. A nostril plugged for more than eight hours meant a serious illness was imminent. If breathing was one-sided for more than a day, death was expected. But why? Ronald Eccles, "A Role for the Nasal Cycle in Respiratory Defense," *European Respiratory Journal* 9, no. 2 (Feb. 1996): 371–76; Eccles et al., "Changes in the Amplitude of the Nasal Cycle Associated with Symptoms of Acute Upper Respiratory Tract Infection," *Acta Otolaryngologica* 116, no. 1 (Jan. 1996): 77–81.

41 **feed more blood to the opposite:** Kahana-Zweig et al.; Shirley Telles et al., "Alternate-Nostril Yoga Breathing Reduced Blood Pressure While Increasing Performance in a Vigilance Test," *Medical Science Monitor Basic Research* 23 (Dec. 2017): 392–98; Karamjit Singh et al., "Effect of Uninostril Yoga Breathing on Brain Hemodynamics: A Functional Near-Infrared Spectroscopy Study," *International Journal of Yoga* 9, no. 1 (June 2016): 12–19; Gopal Krushna Pal et al., "Slow Yogic Breathing Through Right and Left Nostril Influences Sympathovagal Balance, Heart Rate Variability, and Cardiovascular Risks in Young Adults," *North American Journal of Medical Sciences* 6, no. 3 (Mar. 2014): 145–51.

42 **lowers temperature and blood pressure:** P. Raghuraj and Shirley Telles, "Immediate Effect of Specific Nostril Manipulating Yoga Breathing Practices on Autonomic and Respiratory Variables," *Applied Psychophysiology and Biofeedback* 33, no. 2 (June 2008): 65–75. S. Kalaivani, M. J. Kumari, and G. K. Pal, "Effect of Alternate Nostril Breathing Exercise on Blood Pressure, Heart Rate, and Rate Pressure Product among Patients with Hypertension in JIPMER, Puducherry," *Journal of Education and Health Promotion* 8, no. 145 (July 2019).

42 **negative emotions:** Neuroanatomist Jill Bolte Taylor offers an emotional and astonishing primer of the functions of right and left brain in her 2008 TED Talk, "My Stroke of Insight," which, as of this writing, has been viewed more than 26 million times. View it here: https://www.ted.com/talks/jill_bolte_taylor_s_powerful_stroke_of_insight?language=en.

42 **researchers at the University of California:** David Shannahoff-Khalsa and Shahrokh Golshan, "Nasal Cycle Dominance and Hallucinations in an Adult Schizophrenic Female," *Psychiatry Research* 226, no. 1 (Mar. 2015): 289–94.

42 **alternate nostril breathing:** Studies conducted at research labs and published in the *International Journal of Neuroscience*, *Frontiers in Neural Circuits*, *Journal of Laryngology and Otology*, and more have demonstrated clear links between right and left nostrils and specific biological and mental functions. You can find several dozens of studies here: https://www.ncbi.nlm.nih.gov/pubmed/?term=alternate+nostril+breathing.

43 **heat up my body and aid my digestion:** When yogis finish a meal, they lie on their left side so that they will breathe primarily from their right nostril. The increase of blood flow and heat via right-nostril breathing, yogis believe, can aid in digestion. A few years ago, researchers at Jefferson Medical College in Philadelphia tested this claim by feeding 20 healthy subjects a high-fat meal on different days, having them lie on their right or left sides. Those ordered to lie on their left side (breathing primarily through their right nostril) had significantly less heartburn and measured much lower acidity in their throats than subjects lying on the right side. The study was repeated with the same results. The extra heating in the body triggered by right nostril breathing likely influenced the rate and efficiency of digestion, but

gravity certainly helped. The stomach and pancreas will hang more naturally when the body is positioned on the left side, which allows food to more easily move through the large intestine. In short, it feels better and is more efficient for digestion. L. C. Katz et al., "Body Position Affects Recumbent Postprandial Reflux," *Journal of Clinical Gastroenterology* 18, no. 4 (June 1994): 280–83; Anahad O'Connor, "The Claim: Lying on Your Left Side Eases Heartburn," *The New York Times*, Oct. 25, 2010, https://www.nytimes.com /2010/10/26/health/26really.html; R. M. Khoury et al., "Influence of Spontaneous Sleep Positions on Nighttime Recumbent Reflux in Patients with Gastroesophageal Reflux Disease," *American Journal of Gastroenterology* 94, no. 8 (Aug. 1999): 2069–73.

44 all the grains: All the world's beaches contain somewhere in the neighborhood of 2.5 to 10 sextillion grains of sand. Meanwhile, that breath of air you just inhaled contains around 25 sextillion molecules. Fraser Cain, "Are There More Grains of Sand Than Stars?," Universe Today, Nov. 25, 2013, https://www.universetoday.com /106725/are-there-more-grains-of-sand -than-stars/.

44 keep invaders out: Wait, there's more. Katherine J. Wu, "Doctors May Have Found Secretive New Organs in the Center of Your Head," *The New York Times*, Oct. 19, 2020, https://nytimes.com/2020/10/19/ health/saliva-glands-new-organs.html.

45 "first line of defense": "Mucus: The First Line of Defense," ScienceDaily, Nov. 6, 2015, https://www.sciencedaily.com/re leases/2015/11/151106062716.htm; Sara G. Miller, "Where Does All My Snot Come From?," Live Science, May 13, 2016, https:// www.livescience.com/54745 -why-do-i-have-so-much-snot.html; B. M. Yergin et al., "A Roentgenographic Method for Measuring Nasal Mucous Velocity," *Journal of Applied Physiology: Respiratory, Environmental and Exercise Physiology* 44, no. 6 (June 1978): 964–68.

45 tiny, hair-like structures: Maria Carolina Romanelli et al., "Nasal Ciliary Motility: A New Tool in Estimating the Time of Death," *International Journal of*

Legal Medicine 126, no. 3 (May 2012): 427–33; Fuad M. Baroody, "How Nasal Function Influences the Eyes, Ears, Sinuses, and Lungs," *Proceedings of the American Thoracic Society* 8, no. 1 (Mar. 2011): 53–61; Irina Ozerskaya et al., "Ciliary Motility of Nasal Epithelium in Children with Asthma and Allergic Rhinitis," *European Respiratory Journal* 50, suppl. 61 (2017).

45 16 beats per second: The hotter it is, the faster cilia move. J. Yager et al., "Measurement of Frequency of Ciliary Beats of Human Respiratory Epithelium," *Chest* 73, no. 5 (May 1978): 627–33; James Gray, "The Mechanism of Ciliary Movement. VI. Photographic and Stroboscopic Analysis of Ciliary Movement," *Proceedings of the Royal Society B: Biological Sciences* 107, no. 751 (Dec. 1930): 313–32.

45 Cilia closer to the nostrils: Crying will drain tears into the nose, which mix with the mucus, making it thin and watery. Cilia can no longer hold onto the mucus, so it starts running with the flow of gravity: a runny nose. Thick mucus is worse. Excessive dairy, allergies, starchy foods, and more increase mucus weight and density. Cilia slow down, become overwhelmed, and eventually come to a dead stop. This is how the nose gets congested. The longer the nose is stopped up, the more microbes build up, resulting sometimes in a nasal infection (sinusitis) or a common cold. Olga V. Plotnikova et al., "Primary Cilia and the Cell Cycle," *Methods in Cell Biology* 94 (2009): 137–60; Achim G. Beule, "Physiology and Pathophysiology of Respiratory Mucosa of the Nose and the Paranasal Sinuses," *GMS Current Topics in Otorhinolaryngology– Head and Neck Surgery* 9 (2010): Doc07.

45 turbinates will heat: Scheithauer, "Surgery of the Turbinates," 18; Swami Rama, Rudolph Ballentine, and Alan Hymes, *Science of Breath: A Practical Guide* (Honesdale, PA: Himalayan Institute Press, 1979, 1998), 45.

45 Around 1500 BCE: Bryan Gandevia, "The Breath of Life: An Essay on the Earliest History of Respiration: Part I," *Australian Journal of Physiotherapy* 16, no. 1 (Mar. 1970): 5–11, https://www.sciencedirect

.com/science/article/pii/S0004951414610 850; Gandevia, "The Breath of Life: An Essay on the Earliest History of Respiration: Part II," *Australian Journal of Physiotherapy* 16, no. 2 (June 1970): 57–69, https://www.sciencedirect.com/science/article/pii/S0004951414610898?via%3Dihub.

46 a portrait painter: The following details, quotes, and descriptions about George Catlin are taken from the following books and writings: George Catlin, *North American Indians,* ed. Peter Matthiessen (New York: Penguin, 2004); Catlin, *The Breath of Life,* 4th ed., retitled *Shut Your Mouth and Save Your Life* (London: N. Truebner, 1870). The 1870 edition of *Shut Your Mouth* may be read and downloaded for free at https://buteykoclinic.com/wp-content/uploads/2019/04/Shut-your-mouth-Catlin.pdf.

46 "I am traveling": Catlin, *Letters and Notes on the Manners, Customs, and Condition of the North American Indians* (New York: Wiley and Putnam, 1841), vol. 1, 206.

47 "the first, last, and only": Peter Matthiessen, introduction to Catlin, *North American Indians,* vi.

47 all 50 tribes: Later, anthropologist Richard Steckel confirmed Catlin's descriptions, claiming that the Plains tribes in the late 1800s were the tallest people on Earth at the time. Devon Abbot Mihesuah, *Recovering Our Ancestors' Gardens* (Lincoln: University of Nebraska Press, 2005), 47.

47 teeth that were perfectly straight: *Shut Your Mouth,* 2, 18, 27, 41, 43, 51.

48 *The Breath of Life:* Reviewed in *Littell's Living Age* 72 (Jan.–Mar. 1862): 334–35.

49 to be 76: By the 1900s, Catlin was all but forgotten. His mentors, the great Plains Indians, were all but destroyed: killed off by smallpox, shot, raped, or enslaved. The few left often turned to alcohol. The silver-haired Mandan, the broad-shouldered Pawnee, the gentle Minatree—all gone. And with them, so vanished their knowledge of the art and science of breathing.

49 breathe through the nose: Decades after Catlin's treatise on all things mouth and nasal breathing, the physician-in-charge at Mount Regis Sanatorium in Salem, Virginia, a man named E. E. Watson, announced at the annual meeting of the Medical Society of Virginia that mouth-breathing was the primary culprit in the spread of tuberculosis. "To say that seventy-five per cent of our unquestioned tuberculous larynx cases have occurred in mouth-breathers would be no exaggeration," announced Watson. Respiratory disease didn't afflict populations randomly, and they weren't genetic. What Watson was saying, essentially, was that some diseases were a choice. And health or sickness was determined in large part by whether his patients breathed from the mouth or from the nose. E. E. Watson, "Mouth-Breathing," *Virginia Medical Monthly* 47, no. 9 (Dec. 1920): 407–8.

49 written a book: Mark Burhenne, *The 8-Hour Sleep Paradox: How We Are Sleeping Our Way to Fatigue, Disease and Unhappiness* (Sunnyvale, CA: Ask the Dentist, 2015).

49 mouthbreathing contributed: J. E. Choi et al., "Intraoral pH and Temperature during Sleep with and without Mouth Breathing," *Journal of Oral Rehabilitation* 43, no. 5 (Dec. 2015): 356–63; Shirley Gutkowski, "Mouth Breathing for Dummies," *RDH Magazine,* Feb. 13, 2015, https://www.rdhmag.com/patient-care/article/16405394/mouth-breathing-for-dummies.

49 for a hundred years: "Breathing through the Mouth a Cause of Decay of the Teeth," *American Journal of Dental Science* 24, no. 3 (July 1890): 142–43, https://www.ncbi.nlm.nih.gov/pmc/articles/PMC6063589/?page=1.

50 contributor to snoring: M. F. Fitzpatrick et al., "Effect of Nasal or Oral Breathing Route on Upper Airway Resistance During Sleep," *European Respiratory Journal* 22, no. 5 (Nov. 2003): 827–32.

50 a huge boost: To many researchers, nitric oxide is as essential to the body as oxygen and carbon dioxide. Catharine Paddock, "Study Shows Blood Cells Need Nitric Oxide to Deliver Oxygen," *Medical News Today,* Apr. 13, 2015, https://www.medicalnewstoday.com/articles/292292.php; J. Lundberg and E. Weitzberg, "Nasal Nitric Oxide in Man," *Thorax* 54, no. 10 (Oct. 1999): 947–52.

50 **18 percent more oxygen:** J. Lundberg, "Nasal and Oral Contribution to Inhaled and Exhaled Nitric Oxide: A Study in Tracheotomized Patients," *European Respiratory Journal* 19, no. 5 (2002): 859–64; Mark Burhenne, "Mouth Taping: End Mouth Breathing for Better Sleep and a Healthier Mouth," Ask the Dentist (includes several study references), https://askthedentist.com/mouth-tape-better-sleep/. Additionally, the increased air resistance through nasal breathing increases the vacuum in the lungs, and helps us draw in 20 percent more oxygen than through the mouth. Caroline Williams, "How to Breathe Your Way to Better Memory and Sleep," *New Scientist*, Jan. 8, 2020.

51 **my own experiments:** Sleep tape has its critics. A *Guardian* newspaper story from July 2019 claimed that sleep taping was dangerous because "if you started vomiting there would be a good chance you would choke." This claim, Burhenne and Kearney told me, is as ridiculous as it is unfounded and under-researched. "Buteyko: The Dangerous Truth about the New Celebrity Breathing Sensation," *The Guardian*, https://www.theguardian.com/lifeandstyle/shortcuts/2019/jul/15/buteyko-the-dangerous-truth-about-the-new-celebrity-breathing-sensation.

Chapter Four: Exhale

53 **known as a lover:** Publisher's introduction to Peter Kelder, *Ancient Secret of the Fountain of Youth*, Book 2 (New York: Doubleday, 1998), xvi.

54 **lung-expanding stretches:** The directions I followed were on Wikipedia, "Five Tibetan Rites." Cardiologist Joel Kahn suggests performing each rite for 21 rounds, as the ancient Tibetans did. For beginners, ten minutes a day for all the exercises is a good starting point.

54 **extend life:** A half century later, Kelder's booklet was rereleased as *Ancient Secret of the Fountain of Youth*. It became an international sensation, selling more than two million copies. A review of some of the cardiopulmonary benefits of practicing the Five Tibetan Rites can be found in an article by Dr. Joel Kahn, "A Cardiologist's Favorite Yoga Sequence for Boosting Heart Health," MindBodyGreen, Sept. 10, 2019.

55 **according to the researchers:** W. B. Kannel et al., "Vital Capacity as a Predictor of Cardiovascular Disease: The Framingham Study," *American Heart Journal* 105, no. 2 (Feb. 1983): 311–15; William B. Kannel and Helen Hubert, "Vital Capacity as a Biomarker of Aging," in *Biological Markers of Aging*, ed. Mitchell E. Reff and Edward L. Schneider, NIH Publication no. 82-2221, Apr. 1982, 145–60.

55 **comparing lung capacity:** Holgar Shunemann, the researcher who headed the Buffalo follow-up study, reported: "It is important to note that the risk of death was increased for participants with moderately impaired lung function, not merely those in the lowest quintile. This suggests that the increased risk isn't confined to a small fraction of the population with severely impaired lung function." Lois Baker, "Lung Function May Predict Long Life or Early Death," University at Buffalo News Center, Sept. 12, 2000, http://www.buffalo.edu/news/releases/2000/09/4857.html.

55 **results were the same:** The lung-size metric extended to those with lung transplants. In 2013, Johns Hopkins researchers compared several thousand patients who had had lung transplants and found that those who received oversized lungs had a 30 percent increased chance of survival a year after the operation. "For lung transplant, researchers surprised to learn bigger appears to be better," ScienceDaily, Aug. 1, 2013, https://www.sciencedaily.com/releases/2013/08/130801095507.htm; Michael Eberlein et al., "Lung Size Mismatch and Survival After Single and Bilateral Lung Transplantation," *Annals of Thoracic Surgery* 96, no. 2 (Aug. 2013): 457–63.

56 **14 liters:** Brian Palmer, "How Long Can You Hold Your Breath?," *Slate*, Nov. 18, 2013, https://slate.com/technology/2013/11/nicholas-mevoli-freediving-death-what-happens-to-people-who-practice-holding-their-breath.html; https://www.sciencedaily.com/releases/2013/08/130801095507.htm; "Natural Lung Function Decline vs. Lung Function Decline with COPD," *Exhale*, the official blog of

the Lung Institute, Apr. 27, 2016, https://lunginstitute.com/blog/natural-lung-function-decline-vs-lung-function-decline-with-copd/.

56 **15 percent:** I've been asked by more than one musician over the past few years if playing wind instruments boosts lung capacity. Some studies conflict, but the consensus is that, no, wind instruments do not increase lung capacity in a significant way. Further, the constant pressurized air within the lungs appears to increase the risk of chronic upper airway symptoms and even lung cancer. Evangelos Bouros et al., "Respiratory Function in Wind Instrument Players," *Mater Sociomedica* 30, no. 3 (Oct. 2018): 204–8; E. Zuskin et al., "Respiratory Function in Wind Instrument Players," *La Medicina del Lavoro*, Mar. 2009; 100(2); 133–141; A. Ruano-Ravina et al., "Musicians Playing Wind Instruments and Risk of Lung Cancer: Is There an Association?," *Occupational and Environmental Medicine* 60, no. 2 (Feb. 2003); "How to Increase Lung Capacity in 5 Easy Steps," *Exhale*, July 27, 2016.

56 **Katharina Schroth:** Descriptions and details about Schroth and her work were adapted from Hans-Rudolf Weiss, "The Method of Katharina Schroth—History, Principles and Current Development," *Scoliosis and Spinal Disorders* 6, no. 1 (Aug. 2011): 17.

60 **became so renowned:** Descriptions, quotes, and other information regarding Carl Stough and his methods were taken from his 1970 autobiography, coauthored with Reece Stough: *Dr. Breath: The Story of Breathing Coordination* (New York: William Morrow, 1970), 17, 19, 38, 42, 66, 71, 83, 86, 93, 101, 111, 117, 113, 156, 173; a short bio, "Carl Stough," at www.breathingcoordination.ch/en/method/carl-stough; and the documentary film written by Laurence A. Caso, *Breathing: The Source of Life*, Stough Institute, 1997.

61 **making the condition worse:** This was the same "chest" breathing Stough would see in schizophrenics and others with behavioral disorders. They all shared the same tight chest and rib cage, and were unable to move freely or breathe any way but in several hasty breaths. As a result, all the "stale" carbon-dioxide-rich air would sit stagnant in their lungs, creating "dead space."

61 **couldn't get enough stale air out:** In each exhale, we expel about 3,500 compounds. Much of this is organic (water vapor, carbon dioxide, and other gases), but we also exhale pollutants: pesticides, chemicals, and engine exhaust. When we don't breathe out completely, these toxins sit in the lungs and fester, causing infections and other problems. Todor A. Popov, "Human Exhaled Breath Analysis," *Annals of Allergy, Asthma & Immunology* 106, no. 6 (June 2011): 451–56; Joachim D. Pleil, "Breath Biomarkers in Toxicology," *Archives of Toxicology* 90, no. 11 (Nov. 2016): 2669–82; Jamie Eske, "Natural Ways to Cleanse Your Lungs," *Medical News Today*, Feb. 18, 2019, https://www.medicalnewstoday.com/articles/324483.php.

61 **once a minute:** "How Quickly Does a Blood Cell Circulate?," The Naked Scientists, Apr. 29, 2012, https://www.thenakedscientists.com/articles/questions/how-quickly-does-blood-cell-circulate.

61 **2,000 gallons of blood:** "How the Lungs Get the Job Done," American Lung Association, July 20, 2017, https://www.lung.org/about-us/blog/2017/07/how-your-lungs-work.html.

62 **"the second heart":** An overview of Stephen Elliott's theories and observations on the thoracic pump can be found at Stephen Elliot, "Diaphragm Mediates Action of Autonomic and Enteric Nervous Systems," *BMED Reports*, Jan. 8, 2010, https://www.bmedreport.com/archives/8309; also see "Principles of Breathing Coordination" summarized at Breathing Coordination, http://www.breathingcoordination.com/Principles.html.

63 **said Dr. Robert Nims:** Caso, *Breathing: The Source of Life*, 17:12.

64 **asthma, and other respiratory problems:** And the risk of asthma, which in turn affects cardiovascular health. "Adults Who Develop Asthma May Have Higher Risk of Heart Disease, Stroke," *American Heart Association News*, Aug. 24, 2016, https://newsarchive.heart.org/adults-who-develop-asthma-may-have-higher-risk-of-heart-disease-stroke; A. Chaouat

et al., "Pulmonary Hypertension in COPD," *European Respiratory Journal* 32, no. 5 (Nov. 2008): 1371–85.

64 **was relatively easy:** When muscles in the body get strained, other muscles in the area step in to lighten the load. Should we strain our left ankle, we'll place more weight on the right. But the diaphragm doesn't have that option. No other muscle does what it does. It just keeps laboring on at whatever cost, because if it doesn't, we'll quickly run out of air and die. Over time, the body learns to do what it can to compensate and engages "accessory" respiratory muscles in the chest to help air get in and out of the lungs. This chest-centered breathing becomes a habit.

66 **said Lee Evans:** Caso, *Breathing: The Source of Life*, 11:18.

66 **greatest performances:** Bob Burns, *The Track in the Forest: The Creation of a Legendary 1968 US Olympic Team* (Chicago: Chicago Review Press, 2018); Richard Rothschild, "Focus Falls Again on '68 Olympic Track Team," *Chicago Tribune*, June 19, 1998.

66 **power of harnessing:** Along my journey researching this book, I visited Dr. J. Tod Olin, a pulmonologist at National Jewish Health, a leading respiratory hospital and research center in Denver, Colorado. Olin had specialized for the past several years in a condition called exercise-induced laryngeal obstruction (EILO), in which the vocal cords and surrounding structures obstruct the airway during high-intensity exercise. Five to 10 percent of the adolescent population shares this condition, and it is most often misdiagnosed as asthma and treated as such without success. Olin's techniques, which he unimaginatively named Olin EILOBI (Exercise-Induced Laryngeal Obstruction Biphasic Inspiration Techniques), involved restricted and pursed-lip breathing exercises developed by Konstantin Buteyko 60 years earlier, and to a lesser extent, Stough. The only difference was that Olin's techniques were focused through the mouth, because, he said, athletes couldn't inhale fast enough through their noses during high-intensity exercises. One wonders how any of them would have fared if they could. Sarah Graham et al., "The Fortuitous Discovery of the Olin EILOBI Breathing Techniques: A Case Study," *Journal of Voice* 32, no. 6 (Nov. 2018): 695–97.

68 **nearly 4 million Americans:** "Chronic Obstructive Pulmonary Disease (COPD)," Centers for Disease Control and Prevention, National Health Interview Survey, 2018, https://www.cdc.gov/nchs/fastats /copd.htm; "Emphysema: Diagnosis and Treatment," Mayo Clinic, Apr. 28, 2017, https://www.mayoclinic.org/diseases -conditions/emphysema/diagnosis -treatment/drc-20355561.

Chapter Five: Slow

70 **100 times more:** John N. Maina, "Comparative Respiratory Physiology: The Fundamental Mechanisms and the Functional Designs of the Gas Exchangers," *Open Access Animal Physiology* 2014, no. 6 (Dec. 2014): 53–66, https://www.dovepress.com /comparative-respiratory-physiology-the -fundamental-mechanisms-and-the-- peer-reviewed-fulltext-article-OAAP.

72 **"Until the seventeenth century":** Richard Petersham; Campbell, *The Respiratory Muscles and the Mechanics of Breathing*.

73 **more than 1,500 miles:** "How Your Lungs Get the Job Done," American Lung Association, July 2017, https://www.lung.org /about-us/blog/2017/07/how-your-lungs -work.html.

74 **begin a return journey:** Each blood cell offloads only about 25 percent of the oxygen; the remaining 75 percent stays on board and goes back to the lungs. The oxygen that doesn't get off is considered a reserve mechanism, but if the hemoglobin doesn't pick up new oxygen in the lungs, it will be essentially totally empty after about three circulations, which takes about three minutes.

74 **appearance of blood:** "Why Do Many Think Human Blood Is Sometimes Blue?," NPR, Feb. 3, 2017, https://www.npr.org /sections/13.7/2017/02/03/513003105 /why-do-many-think-human-blood -is-sometimes-blue.

74 **the body loses weight:** Ruben Meerman and Andrew J. Brown, "When Somebody Loses Weight, Where Does the Fat Go?," *British Medical Journal* 349 (Dec. 2014): g7257; Rachel Feltman and Sarah Kaplan,

"Dear Science: When You Lose Weight, Where Does It Actually Go?," *The Washington Post*, June 6, 2016.

75 **By his early 30s, Bohr:** If that last name sounds familiar, it should. Christian Bohr was the father of the famed quantum physicist and Nobel laureate Niels Bohr.

75 **Bohr gathered chickens:** L. I. Irzhak, "Christian Bohr (On the Occasion of the 150th Anniversary of His Birth)," *Human Physiology* 31, no. 3 (May 2005): 366–68; Paulo Almeida, *Proteins: Concepts in Biochemistry* (New York: Garland Science, 2016), 289.

76 **to separate oxygen:** Albert Gjedde, "Diffusive Insights: On the Disagreement of Christian Bohr and August Krogh at the Centennial of the Seven Little Devils," *Advances in Physiology Education* 34, no. 4 (Dec. 2010): 174–85.

76 **This discovery explained:** And, of course, the shift in the oxyhemoglobin disassociation curve, the graph which described the relation between partial pressure of oxygen and oxygen saturation of hemoglobin.

76 **Bohr published a paper:** An HTML version is available at https://www1.udel.edu /chem/white/C342/Bohr(1904).html.

76 **Yandell Henderson:** John B. West, "Yandell Henderson," in *Biographical Memoirs*, vol. 74 (Washington, DC: National Academies Press, 1998), 144–59, https://www .nap.edu/read/6201/chapter/9.

76 **"Although clinicians":** Yandell Henderson, "Carbon Dioxide," *Cyclopedia of Medicine*, vol. 3 (Philadelphia: F. A. Davis, 1940). (Several sources list the date as both 1940 and 1934; it's likely the article appeared in both editions.) Lewis S. Coleman, "Four Forgotten Giants of Anesthesia History," *Journal of Anesthesia and Surgery* 3, no. 2 (Jan. 2016): 1–17; Henderson, "Physiological Regulation of the Acid-Base Balance of the Blood and Some Related Functions," *Physiological Reviews* 5, no. 2 (Apr. 1925): 131–60.

77 **of no benefit:** This post sums it up nicely with several quotations from researchers in the field: John A. Daller, MD, "Oxygen Bars: Is a Breath of Fresh Air Worth It?," *On Health*, June 22, 2017, https://www .onhealth.com/content/1/oxygen _bars_-_is_a_breath_of_fresh_air

_worth_it. Additional context can be found in this heavy tome: Nick Lane, *Oxygen: The Molecule That Made the World* (New York: Oxford University Press), 11.

77 **awful experiments:** Yandell Henderson, "Acapnia and Shock. I. Carbon-Dioxid [*sic*] as a Factor in the Regulation of the Heart-Rate," *American Journal of Physiology* 21, no. 1 (Feb. 1908): 126–56.

81 **"of the word *fitness*":** John Douillard, *Body, Mind, and Sport: The Mind-Body Guide to Lifelong Health, Fitness, and Your Personal Best*, rev. ed. (New York: Three Rivers Press, 2001), 153, 156, 211.

81 **On the second day:** I should note that on the first day I switched from mouth-breathing to these slow, nasal breaths, my performance suffered: a .44-mile drop in distance compared to my best mouth-breathing performance a week before. This was to be expected. Conditioning the body to constant, slower nasal breathing takes time. Douillard warned his athletes that they should be prepared for a 50 percent decrease in performance after they first switched to nasal breathing. Some athletes had to wait several months to see gains, which is one reason so many of them, and other non-athletes, give up and just return to mouthbreathing. It's also important to note that these kinds of long inhales and exhales aren't beneficial, or even possible, for very high intensity exercise. Running 400 meters, for instance, would require much more oxygen to keep up with metabolic needs. Some elite athletes can breathe 200 liters of breath per minute during moments of extreme stress—that's up to 20 times what's considered a normal resting volume. But for steady, medium-level exercise like bike riding or jogging, long breaths are far more efficient.

82 **Japanese, African, Hawaiian:** Meryl Davids Landau, "This Breathing Exercise Can Calm You Down in a Few Minutes," *Vice*, Mar. 16, 2018; Christophe André, "Proper Breathing Brings Better Health," *Scientific American*, Jan. 15, 2019.

83 **Ave Maria:** Luciano Bernardi et al., "Effect of Rosary Prayer and Yoga Mantras on Autonomic Cardiovascular Rhythms: Comparative Study," *British Medical Journal* 323, no. 7327 (Dec. 2001): 144649;

T. M. Srinivasan, "Entrainment and Coherence in Biology," *International Journal of Yoga* 8, no. 1 (June 2015): 1–2.

83 **a state of coherence:** Coherence is the measurement of the harmony of two signals. Whenever two signals increase and decrease in phase, they are in coherence, a state of peak efficiency. Much more about coherence and the benefits of breathing 5.5 times a minute with 5.5-second inhales and exhales can be found in the following: Stephen B. Elliott, *The New Science of Breath* (Coherence, 2005); Stephen Elliott and Dee Edmonson, *Coherent Breathing: The Definitive Method* (Coherence, 2008); I. M. Lin, L. Y. Tai, and S. Y. Fan, "Breathing at a Rate of 5.5 Breaths per Minute with Equal Inhalation-to-Exhalation Ratio Increases Heart Rate Variability," *International Journal of Psychophysiolology* 91 (2014): 206–11.

83 **peak efficiency:** A good, doctor-reviewed overview of this type of paced "coherent" breathing: Arlin Cuncic, "An Overview of Coherent Breathing," VeryWellMind, June 25, 2019, https://www.verywellmind .com/an-overview-of-coherent -breathing-4178943.

83 **5.5-second inhales:** 5.4545 breaths a minute, to be exact.

83 **results were profound:** Richard P. Brown and Patricia L. Gerbarg, *The Healing Power of the Breath: Simple Techniques to Reduce Stress and Anxiety, Enhance Concentration, and Balance Your Emotions* (Boston: Shambhala, 2012), Kindle locations 244–47, 1091–96; Lesley Alderman, "Breathe. Exhale. Repeat: The Benefits of Controlled Breathing," *The New York Times*, Nov. 9, 2016.

84 **required no real effort:** In 2012, Italian researchers found that breathing at six breaths a minute had powerful effects at high altitudes of 17,000 feet. The technique not only significantly reduced blood pressure but also boosted oxygen saturation in the blood. Grzegorz Bilo et al., "Effects of Slow Deep Breathing at High Altitude on Oxygen Saturation, Pulmonary and Systemic Hemodynamics," *PLoS One* 7, no. 11 (Nov. 2012): e49074.

84 **"Nobody knows you're doing it":** Landau, "This Breathing Exercise Can Calm You Down."

84 **were in the range of 5.5:** Marc A. Russo et al., "The Physiological Effects of Slow Breathing in the Healthy Human," *Breathe* 13, no. 4 (Dec. 2017): 298–309.

Chapter Six: Less

85 **From around 1850 to 1960:** "Obesity and Overweight," Centers for Disease Control and Prevention, https://www.cdc.gov /nchs/fastats/obesity-overweight.htm; "Obesity Increase," *Health & Medicine*, Mar. 18, 2013; "Calculate Your Body Mass Index," National Heart, Lung, and Blood Institute, https://www.nhlbi.nih.gov /health/educational/lose_wt/BMI/bmi calc.htm?source=quickfitnesssolutions.

85 **offers a troubling picture:** The breathing rate for an average male, according to a study in the 1930s, used to be about 13 times a minute for a total of 5.25 liters of air. By the 1940s, the rate of breathing hovered a bit over 10 breaths a minute for a total of 8 liters. By the 1980s and 1990s, several studies placed the mean breathing rate at closer to 10 to 12 breaths per minute, with a total volume, in some cases, that rose to 9 liters and higher. I discussed this with Dr. Don Storey, a prominent pulmonologist who'd worked in the field for more than 40 years (and who is my father-in-law). He told me that when he was first starting out, the normal respiratory rate was about 8 to 12 breaths a minute. The high end of that rate is nearly doubled today. Beyond the anecdotes, dozens of studies suggest we could indeed be breathing more than we used to. Most studies compare subjects with respiratory illnesses against healthy controls. It's the data from the healthy controls that was used for this assessment. Several studies were discovered in Artour Rakhimov's book *Breathing Slower and Less: The Greatest Health Discovery Ever* (self-published, 2014). Those studies that could be independently verified were included. I will continue to gather research in this area and post on my website—mrjamesnestor.com/breath. In the meantime, here are several studies:

N. W. Shock and M. H. Soley, "Average Values for Basal Respiratory Functions in Adolescents and Adults," *Journal of Nutrition* 18 (1939): 143–53; Harl W. Matheson and John S. Gray, "Ventilatory Function Tests. III. Resting Ventilation, Metabolism, and Derived Measures," *Journal of Clinical Investigation* 29, no. 6 (1950): 688–92; John Kassabian et al., "Respiratory Center Output and Ventilatory Timing in Patients with Acute Airway (Asthma) and Alveolar (Pneumonia) Disease," *Chest* 81, no. 5 (May 1982): 536–43; J. E. Clague et al., "Respiratory Effort Perception at Rest and during Carbon Dioxide Rebreathing in Patients with Dystrophia Myotonica," *Thorax* 49, no. 3 (Mar. 1994): 240–44; A. Dahan et al., "Halothane Affects Ventilatory after Discharge in Humans," *British Journal of Anaesthesia* 74, no. 5 (May 1995): 544–48; N. E. L. Meessen et al., "Breathing Pattern during Bronchial Challenge in Humans," *European Respiratory Journal* 10, no. 5 (May 1997): 1059–63.

86 **quarter of the modern population:** Mary Birch, *Breathe: The 4-Week Breathing Retraining Plan to Relieve Stress, Anxiety and Panic* (Sydney: Hachette Australia, 2019), Kindle locations 228–31. An overview of how poorly we're breathing can be found here: Richard Boulding et al., "Dysfunctional Breathing: A Review of the Literature and Proposal for Classification," *European Respiratory Review* 25, no. 141 (Sept. 2016): 287–94.

86 **Chinese doctors:** Bryan Gandevia, "The Breath of Life: An Essay on the Earliest History of Respiration: Part I," *Australian Journal of Physiotherapy* 16, no. 1 (Mar. 1970): 5–11.

86 **nine and a half breaths per minute:** It's worth mentioning that early Hindus calculated a normal respiratory rate at a much higher rate of 22,636 breaths a day.

88 **"start extending your exhales":** This kind of long inhale and exhale isn't possible for very high intensity exercise. Running 400 meters, for instance, would require much more oxygen to keep up with metabolic needs. (Endurance athletes can breathe 200 liters of breath per minute during moments of extreme stress—that's up to 20 times what's considered a normal resting volume.) But for steady, medium-level exercise like this, long breaths are far more efficient. Maurizio Bussotti et al., "Respiratory Disorders in Endurance Athletes—How Much Do They Really Have to Endure?," *Open Access Journal of Sports Medicine* 2, no. 5 (Apr. 2014): 49.

89 **increases VO$_2$ max:** Using "slow and less" breathing techniques, subjects in an experiment conducted at the Universitas Muhammadiyah Surakarta, Faculty of Health Science in Indonesia, and presented at the 3rd International Conference on Science, Technology, and Humanity (ISETH) in December 2017, showed a significant VO$_2$ max increase over a control group. Dani Fahrizal and Totok Budi Santoso, "The Effect of Buteyko Breathing Technique in Improving Cardiorespiratory Endurance," *2017 ISETH Proceeding Book* (UMS publications), https://pdfs.semanticscholar.org/c2ee/b2d1c0230a76fccdad94e7d97b11b882d217.pdf; Several more study summaries are available at Patrick McKeown, "Oxygen Advantage," https://oxygenadvantage.com/improved-swimming-coordination.

89 **"diagnosed machine disorders":** K. P. Buteyko, ed., *Buteyko Method: Its Application in Medical Practice* (Odessa, Ukraine: Titul, 1991).

89 **shot to 212:** Details from this biography were taken from several sources. "The Life of Konstantin Pavlovich Buteyko," Buteyko Clinic, https://buteykoclinic.com/about-dr-buteyko; "Doctor Konstantin Buteyko," Buteyko.com, http://www.buteyko.com/method/buteyko/index_buteyko.html; "The History of Professor K. P. Buteyko," LearnButeyko.org, http://www.learnbuteyko.org/the-history-of-professor-kp-buteyko; Sergey Altukhov, *Doctor Buteyko's Discovery* (TheBreathingMan, 2009), Kindle locations 570, 572, 617; Buteyko interview, 1988, YouTube, https://www.youtube.com/watch?v=yv5unZd7okw.

92 **headed to Akademgorodok:** "The Original Silicon Valley," *The Guardian*, Jan. 5, 2016, https://www.theguardian.com/artandddesign/gallery/2016/jan/05/akadem

gorodok-academy-town-siberia-science-russia-in-pictures.

92 **Laboratory of Functional Diagnostics:** See an amazing photo of the lab here: https:// images.app.goo.gl/gAHupj GqjBtEiKab9.

93 **6.5 to 7.5 percent carbon dioxide:** A copy of Buteyko's carbon dioxide chart can be found here: https://tinyurl.com/yy3fvrh7.

93 **Buteyko developed a protocol:** Buteyko's papers and musings can be downloaded for free on Patrick McKeown's website: https://tinyurl.com/y3lbfhx2.

94 **Zátopek developed:** More about hypoventilation training is available on Dr. Xavier Woorons's website: http://www.hypoventilation-training.com/index.html; "Emil Zatopek Biography," Biography Online, May 1, 2010, https://www.biographyonline.net/sport/athletics/emile-zatopek.html; Adam B. Ellick, "Emil Zatopek," *Runner's World*, Mar. 1, 2001, https://www.runnersworld.com/advanced/a20841849/emil-zatopek. For what it's worth, Zátopek's height is something of a mystery; some references state he was six feet tall but others, such as ESPN, have him as five-six. The consensus, according to *Runner's World*, is that he was about five-eight.

94 **widely derided:** Timothy Noakes, *Lore of Running*, 4th ed. (Champaign, IL: Human Kinetics, 2002), 382.

94 **would later be named:** "Emil Zátopek," Running Past, http://www.runningpast.com/emil_zatopek.htm; Frank Litsky, "Emil Zatopek, 78, Ungainly Running Star, Dies," *The New York Times*, Nov. 23, 2000, https://www.nytimes.com/2000/11/23/sports/emil-zatopek-78-ungainly-running-star-dies.html.

94 **"hurt, pain, and agony":** Joe Hunsaker, "Doc Counsilman: As I Knew Him," *SwimSwam*, Jan. 12, 2015, https://swimswam.com/doc-counsilman-knew/.

95 **swim faster:** Some interesting context on the possible dangers of Counsilman's approach in training for younger athletes by swim coach Mike Lewellyn: https://swimisca.org/coach-mike-lewellyn-on-breath-holding-shallow-water-blackout/. An alternative view by Dr. Rob Orr can be found at "Hypoxic Work in the Pool," PTontheNet, Feb. 14, 2006, https:// www.ptonthenet.com /articles/ Hypoxic-Work-in-the-Pool -2577. What I've surmised from these and several other posts is that hypoxia training works, but it should not be employed as one-size-fits-all training regimen. Physiological, psychological, and numerous anatomical factors must all be considered, as with most any other, training technique. And like any other underwater training, hypoxic training must always be under the close supervision of professionals.

95 **Counsilman used it:** "ISHOF Honorees," International Swimming Hall of Fame, https:// ishof.org/dr.-james-e.--doc--counsilman-(usa).html; "A Short History: From Zátopek to Now," Hypoventilation Training.com, http://www.hypoventilation-training.com/historical.html.

95 **U.S. Olympic swim team:** Braden Keith, "Which Was the Greatest US Men's Olympic Team Ever?," *SwimSwam*, Sept. 7, 2010, https:// swimswam.com/which-was -the-greatest-us-mens-olympic-team -ever; Jean-Claude Chatard, ed., *Biomechanics and Medicine in Swimming IX* (Saint-Étienne, France: University of Saint-Étienne Publications, 2003).

95 **boost in red blood cells:** To be clear, Woorons's research is directed at elite athletes looking to get an edge on the competition. Nobody knows the long-term effects of consistently pushing the body into a highly anaerobic state, and several researchers suggest such constant anaerobic workouts may break down the body and cause damaging oxidative stress. Meanwhile, with just a few weeks of Olsson's lighter, milder training, several of his clients registered significant gains in red blood cell counts. More blood means more oxygen delivered to more tissues. Lance Armstrong, the disgraced cyclist, didn't get busted for taking adrenaline or steroids but for injecting himself with his own blood and increasing his red blood cell count, which would allow him to carry more oxygen. What Armstrong was essentially doing was a instant fix of breath restriction training.

95 **Breathing *way less*:** Xavier Woorons et al., "Prolonged Expiration down to Residual Volume Leads to Severe Arterial

Hypoxemia in Athletes during Sub-maximal Exercise," *Respiratory Physiology & Neurobiology* 158, no. 1 (Aug. 2007): 75–82; Alex Hutchinson, "Holding Your Breath during Training Can Improve Performance," *The Globe and Mail*, Feb. 23, 2018, https://www.theglobeandmail.com /life/health-and-fitness/fitness/holding -your-breath-during-training-can -improve-performance/article38089753/.

96 **Just a few weeks:** E. Dudnik et al., "Intermittent Hypoxia-Hyperoxia Conditioning Improves Cardiorespiratory Fitness in Older Comorbid Cardiac Outpatients without Hematological Changes: A Randomized Controlled Trial," *High Altitude Medical Biology* 19, no. 4 (Dec. 2018): 339–43. And much more. A British study of 30 rugby players showed those trained in "normobaric" levels of 13 percent oxygen (equivalent of an altitude of 12,000 feet) had "twofold greater improvements" over controls training in normal sea-level air after just four weeks. A European study of 86 obese women showed hypoxia training led to a "significant decrease in waist circumference" and significant reduction in fat over controls. (More available oxygen in the cells meant that more fat could be burned more efficiently.) And even diabetes! Twenty-eight adults suffering from type 1 diabetes showed that hypoxia training reduced glucose concentrations, keeping the subjects more in line with normal levels over controls. The simple method, the researchers wrote, "may induce significant prevention of diabetes cardiovascular complications." See references for all these studies, and more, at mrjamesnestor.com/breath.

96 **This list goes on:** See a photo of Sanya Richards-Ross competing here: https:// tinyurl.com/yyf8tj7m.

96 **our lungs roughly half full:** Throughout our jogging, Olsson and I used a Relaxator, a device Olsson designed to restrict airflow during exhalations and increase the positive pressure on the lungs, which helps them expand and increase space for gas exchange. Breath resistance devices like the Relaxator can help monitor a consistent flow of air and measure the amount of resistance, but they are optional. The most effective technique in hypoventilation

training is to extend exhales and then hold the breath with lungs half-full as long as possible and do it all over again. This can happen anywhere, at any time. The more "air hunger" you create, the more EPO will release from the kidneys, the more red blood cells will release from bone marrow, the more oxygen will upload into your body, the more resilient the body will become, the farther and faster and higher it will go. In the 1990s, Dr. Alison McConnell, a London physiologist and a leading expert on breath training, had cyclists use a resistance device that forced pressure on the inhale. She found that the athletes gained a shocking 33 percent increase in endurance performance after just four weeks. Just five minutes of this training can lower blood pressure by 12 points, about twice what aerobic exercise delivers. Alison McConnell, *Breathe Strong, Perform Better* (Champaign, IL: Human Kinetics, 2011), 59, 61; Lisa Marshall, "Novel 5-Minute Workout Improves Blood Pressure, May Boost Brain Function," Medical Xpress, Apr. 8, 2019, https://medicalxpress .com/news/2019-04-minute-workout -blood-pressure-boost.html; Sarah Sloat, "A New Way of Working Out Takes 5 Minutes and Is as Easy as Breathing," Inverse, Apr. 9, 2019, https://www.inverse.com/arti cle/54740-imst-training-blood-pressure -health.

98 **50 scientific papers:** An exhaustive list of Buteyko's studies and other research, in both English and Russian, are available at the following links provided by Breathe Well Clinic (Dublin, Ireland) and Buteyko Clinic International: http://breathing.ie /clinical-studies-in-russian/; http://breath ing.ie/clinical-evidence-for-buteyko/; https://buteykoclinic.com/wp-content /uploads/2019/04/Dr-Buteykos-Book.pdf.

98 **25 million Americans:** Stephen C. Redd, "Asthma in the United States: Burden and Current Theories," *Environmental Health Perspectives* 110, suppl. 4 (Aug. 2002): 557–60; "Asthma Facts and Figures," Asthma and Allergy Foundation of America, https://www.aafa.org/asthma-facts; "Childhood Asthma," Mayo Clinic, https:// www.mayoclinic.org/diseases-conditions /childhood-asthma/symptoms-causes /syc-20351507.

98 **fourfold increase:** Paul Hannaway, *What to Do When the Doctor Says It's Asthma* (Gloucester, MA: Fair Winds, 2004).

98 **Pollutants, dust, viral infections, cold air:** "Childhood Asthma," Mayo Clinic, https://www.mayoclinic.org/diseases-conditions/childhood-asthma/symptoms-causes/syc-20351507.

98 **asthma can be brought on:** Duncan Keeley and Liesl Osman, "Dysfunctional Breathing and Asthma," *British Medical Journal* 322 (May 2001): 1075; "Exercise-Induced Asthma," Mayo Clinic, https://www.mayoclinic.org/diseases-conditions/exercise-induced-asthma/symptoms-causes/syc-20372300.

98 **exercise-induced asthma:** R. Khajotia, "Exercise-Induced Asthma: Fresh Insights and an Overview," *Malaysian Family Physician* 3, no. 2 (Apr. 2008): 21–24.

99 **worldwide annual market:** "Distribution of Global Respiratory Therapy Market by Condition in 2017–2018 (in Billion U.S. Dollars)," Statista, https://www.statista.com/statistics/312329/worldwide-respiratory-therapy-market-by-condition/.

99 **worsened asthma symptoms:** When a group of medical doctors, professors, and statisticians wanted to know how medicine and procedures actually affected patients, they didn't look up reviews in WebMD. They noticed that the numbers in many studies were funded by private drug companies and the outcomes were either fudged or grossly misleading. So these researchers gathered studies from dozens of different treatments and reanalyzed the data to offer an accurate measurement of the impact of a medicine or therapy. To give a real-life view into how effective drugs and treatments are, the researchers' outcomes estimated the number of patients who need to be treated to have an impact on one person. They called their organization The NNT, a simple statistical concept: "Number Needed to Treat." Since starting in 2010, NNT (https://www.thennt.com) has surveyed more than 275 drugs and therapies in fields ranging from cardiology to endocrinology to dermatology. They rated each of these drugs and therapies on a color scale: green (the therapy or drug has clear benefits); yellow (it is unclear if it has any benefits); red (no benefits); and black (the treatment is more harmful to patients than helpful). They reviewed 48 trials, including tens of thousands of subjects, of a standard asthma treatment: long-acting beta-agonists (LABA) with corticosteroids, an inhaled combination drug treatment with trade names Advair and Symbicort, designed to keep smooth muscles in the airways constantly relaxed. Of the 48 trials represented, 44 were sponsored by the pharmaceutical maker of long-acting beta-agonists, one of the two drugs in combination. This drug was not only approved, but used by likely millions of asthmatics every year. NNT crunched the numbers and found that combination LABAs and steroid inhalers were not only totally ineffective but harmful. Only 1 in 73 asthmatic patients who used the drug reduced their chances of mild to moderate asthma attack. Meanwhile, the drug provoked a severe asthma attack in 1 in 140 people. According to NNT, the drug "seems to have caused an asthma-related death" for 1 in every 1,400 asthmatics. LABAs were equally ineffective for children. More context on this subject: Vassilis Vassilious and Christos S. Zipitis, "Long-Acting Bronchodilators: Time for a Re-think," *Journal of the Royal Society of Medicine* 99, no. 8 (Aug. 2006): 382–83.

99 **David Wiebe:** Jane E. Brody, "A Breathing Technique Offers Help for People with Asthma," *The New York Times*, Nov. 2, 2009, https://www.nytimes.com/2009/11/03/health/03brod.html; "Almost As If I No Longer Have Asthma After Natural Solution," Breathing Center, Apr. 2009, https://www.breathingcenter.com/now-living-almost-as-if-i-no-longer-have-asthma.

99 **asthma and overall health:** Sasha Yakovleva, K. Buteyko, et al., *Breathe to Heal: Break Free from Asthma (Breathing Normalization)* (Breathing Center, 2016), 246; "Buteyko Breathing for Improved Athletic Performance," Buteyko Toronto, http://www.buteykotoronto.com/buteyko-and-fitness.

100 **Sanya Richards-Ross:** "Buteyko and Fitness," Buteyko Toronto, http://www.buteykotoronto.com/buteyko-and-fitness.

100 **They all breathed better:** Thomas Ritz et al., "Controlling Asthma by Training of Capnometry-Assisted Hypoventilation (CATCH) Versus Slow Breathing: A Randomized Controlled Trial," *Chest* 146, no. 5 (Aug. 2014): 1237–47.

100 **"very strange happening":** "Asthma Patients Reduce Symptoms, Improve Lung Function with Shallow Breaths, More Carbon Dioxide," ScienceDaily, Nov. 4, 2014, https://www.sciencedaily.com/releases/2014/11/141104111631.htm.

101 **A half-dozen other clinical trials:** "Effectiveness of a Buteyko-Based Breathing Technique for Asthma Patients," ARCIM Institute—Academic Research in Complementary and Integrative Medicine, 2017, https://clinicaltrials.gov/ct2/show/NCT03098849.

102 **real damage from overbreathing:** It's worth noting that overbreathing can also cause calcium levels to drop in your blood, which can result in numbness and tingling, muscle spasms, cramps, and twitching.

102 **Weeks, months, or years:** If the body is forced to constantly compensate by excreting bicarbonate, levels of this chemical will begin to taper off, and the pH will waver from its optimum functioning at 7.4. John G. Laffey and Brian P. Kavanagh, "Hypocapnia," *New England Journal of Medicine* 347 (July 2002): 46; G. M. Woerlee, "The Magic of Hyperventilation," Anesthesia Problems & Answers, http://www.anesthesiaweb.org/hyperventilation.php.

103 **becomes even more difficult:** Jacob Green and Charles R. Kleeman, "Role of Bone in Regulation of Systemic Acid-Base Balance," *Kidney International* 39, no. 1 (Jan. 1991): 9–26.

103 **stave off further attacks:** "Magnesium Supplements May Benefit People with Asthma," NIH National Center for Complementary and Integrative Health, Feb. 1, 2010, https://nccih.nih.gov/research/results/spotlight/021110.htm

104 **"The yogi's life is not measured":** Andrew Holecek, *Preparing to Die: Practical Advice and Spiritual Wisdom from the Tibetan Buddhist Tradition* (Boston: Snow Lion, 2013). Animal metrics were taken from these studies: "Animal Heartbeats," Every Second, https://everysecond.io/animal

-heartbeats; "The Heart Project," Public Science Lab, http://robdunnlab.com/projects/beats-per-life/; Yogi Cameron Alborzian, "Breathe Less, Live Longer," *The Huffington Post*, Jan. 14, 2010, https://www.huffpost.com/entry/breathe-less-live-longer_b_422923; Mike McRae, "Do We Really Only Get a Certain Number of Heartbeats in a Lifetime? Here's What Science Says," ScienceAlert, Apr. 14, 2018, https://www.sciencealert.com/relationship-between-heart-beat-and-life-expectancy.

Chapter Seven: Chew

107 **Twelve thousand years ago:** "Malocclusion and Dental Crowding Arose 12,000 Years Ago with Earliest Farmers, Study Shows," University College Dublin News, http://www.ucd.ie/news/2015/02FEB15/050215-Malocclusion-and-dental-crowding-arose-12000-years-ago-with-earliest-farmers-study-shows.html; Ron Pinhasi et al., "Incongruity between Affinity Patterns Based on Mandibular and Lower Dental Dimensions following the Transition to Agriculture in the Near East, Anatolia and Europe," *PLoS One* 10, no. 2 (Feb. 2015): e0117301.

107 **first widespread instances of crooked teeth:** Jared Diamond, "The Worst Mistake in the History of the Human Race," *Discover*, May 1987, http://discovermagazine.com/1987/may/02-the-worst-mistake-in-the-history-of-the-human-race; Jared Diamond, *The Third Chimpanzee: The Evolution and Future of the Human Animal* (New York: HarperCollins, 1992).

108 **The dead were downstairs:** Natasha Geiling, "Beneath Paris's City Streets, There's an Empire of Death Waiting for Tourists," Smithsonian.com, Mar. 28, 2014, https://www.smithsonianmag.com/travel/paris-catacombs-180950160; "Catacombes de Paris," Atlas Obscura, https://www.atlasobscura.com/places/catacombes-de-paris.

109 **largest graveyards on Earth:** The largest being Wadi-us-Salaam in Iraq, which contains tens of millions of bodies.

111 **the average Briton:** Gregori Galofré-Vilà, et al., "Heights across the Last 2000 Years in England," University of Oxford, Discussion Papers in Economic and Social History, no. 151, Jan. 2017, 32, https://www

.economics.ox.ac.uk/materials/working _papers/2830/151-final.pdf. C.W., "Did Living Standards Improve during the Industrial Revolution?," *The Economist*, https:// www.economist.com/free-exchange/2013 /09/13/did-living-standards-improve -during-the-industrial-revolution.

111 **teeth removed altogether:** According to a civil servant at the National Health Service, up to the mid-1990s it was common for women to be given vouchers to have all their teeth removed before their 16th or 18th birthday in areas around the northeast of England. Letters, *London Review of Books* 39, no. 14 (July 2017), https://www .lrb.co.uk/v39/n14/letters.

112 **Victorian dentist observed:** Review of J. Sim Wallace, *The Physiology of Oral Hygiene and Recent Research, with Special Reference to Accessory Food Factors and the Incidence of Dental Caries* (London: Ballière, Tindall and Cox, 1929), in *Journal of the American Medical Association* 95, no. 11 (Sept. 1930): 819.

113 **In the 1800s:** I'm talking about Edward Mellanby, a British researcher who would be knighted for his work and would blame our shrinking faces on deficiencies of vitamin D in the modern diet. An American dentist named Percy Howe thought crooked teeth were caused by lack of vitamin C.

113 **"Since we have known":** Earnest A. Hooton, foreword to Weston A. Price, *Nutrition and Physical Degeneration* (New York: Paul B. Hoeber, 1939). "Let us cease pretending that toothbrushes and toothpaste are any more important than shoe brushes and shoe polish. It is store food that has given us store teeth," Hooton wrote in his own book, *Apes, Men, and Morons* (New York: G. P. Putnam's Sons, 1937).

114 **Price found communities:** When Price later examined samples of the bread and cheese from the village of Loetschental in his laboratory in Cleveland, he found that it contained 10 times the amount of vitamins A and D of all the foods in a typical modern American diet at the time. Price researched the dead as well. In Peru, he painstakingly analyzed 1,276 skulls that ranged from a few hundred to a few thousand years old. Not a single skull had any deformity in its dental arches, not one face

was deformed or misshaped. Weston A. Price, *Nutrition and Physical Degeneration*, 8th ed. (Lemon Grove, CA: Price-Pottenger Nutrition Foundation, 2009).

114 **food was wild animals:** The Native Americans whom Price visited in northern Canada had no access to fruits or vegetables during the long winters, and thus no vitamin C. Price noted that they should have all been sick or dead from scurvy, yet they appeared to be in vigorous health. An elder chief described to Price how the tribe would occasionally kill a moose, cut open its back, and pull out two small balls of fat just above the kidneys. They'd cut up these balls and distribute them among the family. Price later discovered that these balls were the adrenal glands, the richest source of vitamin C in all animal and plant tissues.

115 **But some complained:** "Nutrition and Physical Degeneration: A Comparison of Primitive and Modern Diets and Their Effects," *Journal of the American Medical Association* 114, no. 26 (June 1940): 2589, https://jamanetwork.com/journals/jama /article-abstract/1160631?redirect=true.

118 **Balloon sinuplasty:** Nayak was careful to point out that these patients were a highly selected cohort and that patients needed no other procedural treatments over an additional one-year period. He told me that balloon sinuplasty worked for these patients, but wouldn't work for everyone.

118 **the Cottle's maneuver:** Jukka Tikanto and Tapio Pirilä, "Effects of the Cottle's Maneuver on the Nasal Valve as Assessed by Acoustic Rhinometry," *American Journal of Rhinology* 21, no. 4 (July 2007): 456–59.

118 **have a deviated septum:** Shawn Bishop, "If Symptoms Aren't Bothersome, Deviated Septum Usually Doesn't Require Treatment," Mayo Clinic News Network, July 8, 2011, https://newsnetwork.mayoclinic .org/discussion/if-symptoms-arent -bothersome-deviated-septum-usually -doesnt-require-treatment/.

118 **50 percent of us have:** Sanford M. Archer and Arlen D. Meyers, "Turbinate Dysfunction," Medscape, Feb. 13, 2019.

119 **75 percent of his turbinates removed:** Peter's story was particularly heartwrenching. After his surgeries doctors prescribed antidepressants and told him

he just suffered from age-related problems. He spent the next three years learning to construct an elaborate three-dimensional model from X-rays that he would then use to measure something called "Computational Fluid Dynamics." These before-and-after models and the data allowed him to determine the exact changes to airflow velocity, distribution, temperature, pressure, resistance, and humidity levels that were affected by his previous turbinate surgeries. Overall, his nasal cavity was four times larger than what is considered normal or healthy. His nose had lost the ability to properly heat air, and air was moving through it twice as fast as it should. And still, Peter says, a large proportion of the medical community argues that empty nose syndrome is a psychological problem, not a physical one. Read more about Peter's research: http://emptynosesyndromeaerodynamics.com.

119 **contemplated suicide:** The medical community at large viewed empty nose syndrome as a problem of the mind, not the nose. One doctor went so far as to refer to empty nose syndrome in the *Los Angeles Times* as "empty head syndrome"; Aaron Zitner, "Sniffing at Empty Nose Idea," *Los Angeles Times*, May 10, 2001; Cedric Lemogne et al., "Treating Empty Nose Syndrome as a Somatic Symptom Disorder," *General Hospital Psychiatry* 37, no. 3 (May–June 2015): 273.e9–e10; Joel Oliphint, "Is Empty Nose Syndrome Real? And If Not, Why Are People Killing Themselves Over It?," BuzzFeed, Apr. 14, 2016; Yin Lu, "Kill the Doctors," *Global Times*, Nov. 26, 2013, http://www.globaltimes.cn/content/827820.shtml.

119 **"Hundreds of people":** Dozens of researchers have confirmed so much of what Peter and other empty nose syndrome sufferers have reported: that the condition is the result of real, measurable damage that had occurred from nasal surgery. Chengyu Li, Alexander A. Farag, James Leach, Bhakthi Deshpande, et al., "Computational Fluid Dynamics and Trigeminal Sensory Examinations of Empty Nose Syndrome Patients," *Laryngoscope*, June 1, 2018, E176-184; Jennifer Malik et al., "The Cotton

Test Redistributes Nasal Airflow in Patients with Empty Nose Syndrome," *International Forum of Allergy & Rhinology*, Jan. 2020, 539-545.

120 **up to 20 percent:** Oliphint, "Is Empty Nose Syndrome Real?"

120 **linked to obstruction:** Michael L. Gelb, "Airway Centric TMJ Philosophy," *CDA Journal* 42, no. 8 (Aug. 2014): 551–62, https://pdfs.semanticscholar.org/8bc1/8887d39960f9cce328f5c61ee356e11d0c09.pdf.

121 **risk of airway obstruction:** Felix Liao, *Six-Foot Tiger, Three-Foot Cage: Take Charge of Your Health by Taking Charge of Your Mouth* (Carlsbad, CA: Crescendo, 2017), 59.

121 **Friedman tongue position scale:** Rebecca Harvey et al., "Friedman Tongue Position and Cone Beam Computed Tomography in Patients with Obstructive Sleep Apnea," *Laryngoscope Investigative Otolaryngology* 2, no. 5 (Aug. 2017): 320–24; Pippa Wysong, "Treating OSA? Don't Forget the Tongue," *ENTtoday*, Jan. 1, 2008, https://www.enttoday.org/article/treating-osa-dont-forget-the-tongue/.

121 **clog the throat:** An overview of this dilemma can be found on Dr. Eric Kezirian's website: https://sleep-doctor.com/blog/new-research-treating-the-large-tongue-in-sleep-apnea-surgery.

121 **more than 17 inches:** Liza Torborg, "Neck Size One Risk Factor for Obstructive Sleep Apnea," Mayo Clinic, June 20, 2015, https://newsnetwork.mayoclinic.org/discussion/mayo-clinic-q-and-a-neck-size-one-risk-factor-for-obstructive-sleep-apnea/.

121 **Ninety percent of the obstruction:** Gelb, "Airway Centric TMJ Philosophy"; Luqui Chi et al., "Identification of Craniofacial Risk Factors for Obstructive Sleep Apnoea Using Three-Dimensional MRI," *European Respiratory Journal* 38, no. 2 (Aug. 2011): 348–58.

122 **especially effective for children:** Babies who have breathing issues at six months have a 40 percent greater chance of having behavioral issues (including ADHD) starting around age four, according to Gelb. Michael Gelb and Howard Hindin, *Gasp! Airway Health—The Hidden Path to*

Wellness (self-published, 2016), Kindle location 850.

122 **kids with ADHD:** Chai Woodham, "Does Your Child Really Have ADHD?," *U.S. News,* June 20, 2012, https://health.us news.com/health-news/articles/2012/06 /20/does-your-child-really-have-adhd.

122 **problems that come with it:** More on this very expansive and very depressing subject: "Kids Behave and Sleep Better after Tonsillectomy, Study Finds," press release, University of Michigan Health System, Apr. 3, 2006, https://www.eurekalert .org/pub_releases/2006-04/uomh -kba032806.php; Susan L. Garetz, "Adenotonsillectomy for Obstructive Sleep Apnea in Children," UptoDate, Oct. 2019, https://www.uptodate.com/contents/ade notonsillectomy-for-obstructive-sleep -apnea-in-children. It's worth noting, as well, that according to several studies, most mouthbreathing children are also sleep-deprived, and lack of sleep will have a direct impact on growth. Yosh Jefferson, "Mouth Breathing: Adverse Effects on Facial Growth, Health, Academics, and Behavior," *General Dentistry* 58, no. 1 (Jan.–Feb. 2010): 18–25; Carlos Torre and Christian Guilleminault, "Establishment of Nasal Breathing Should Be the Ultimate Goal to Secure Adequate Craniofacial and Airway Development in Children," *Jornal de Pediatria* 94, no. 2 (Mar.–Apr. 2018): 101–3. A study that followed 1,900 children for 15 years found that children with severe snoring, sleep apnea, and other sleep-disordered breathing were twice as likely to become obese compared to children who didn't snore. Those children who had the worst symptoms had a 60 to 100 percent increased risk of obesity. "Short Sleep Duration and Sleep-Related Breathing Problems Increase Obesity Risk in Kids," press release, Albert Einstein College of Medicine, Dec. 11, 2014.

123 **Norman Kingsley:** Sheldon Peck, "Dentist, Artist, Pioneer: Orthodontic Innovator Norman Kingsley and His Rembrandt Portraits," *Journal of the American Dental Association* 143, no. 4 (Apr. 2012): 393–97.

123 **Pierre Robin:** Ib Leth Nielsen, "Guiding Occlusal Development with Functional Appliances," *Australian Orthodontic Journal* 14, no. 3 (Oct. 1996): 133–42; "Functional Appliances," British Orthodontic Society; John C. Bennett, *Orthodontic Management of Uncrowded Class II Division 1 Malocclusion in Children* (St. Louis: Mosby/Elsevier, 2006); "Isolated Pierre Robin sequence," Genetics Home Reference, https://ghr.nlm.nih.gov/condi tion/isolated-pierre-robin-sequence.

124 **retractive orthodontics:** Edward Angle, considered the "father of American orthodontics," was opposed to extracting teeth; meanwhile, his student, Charles H. Tweed, would go on to champion extractions. In the end, Tweed's approach won. Sheldon Peck, "Extractions, Retention and Stability: The Search for Orthodontic Truth," *European Journal of Orthodontics* 39, no. 2 (Apr. 2017): 109–15.

124 **Dr. John Mew:** Mew had spent three years as a facial surgeon at Queen Victoria Hospital in West Sussex studying how the mouth worked. He knew that the 14 jigsaw-puzzle bones that made up the face needed to develop together in just the right way; any disruption of any of these bones could affect the function and growth of the entire mouth and face.

124 **who'd gotten extractions:** That tooth extractions cause a flattening of the face is not widely accepted in the orthodontics industry. Several studies have claimed that extractions cause retrognathic facial growth, while others show little or no change in the face. Still others say results will vary and can only be determined by first considering the width of the palate. Antônio Carlos de Oliveira Ruellas et al., "Tooth Extraction in Orthodontics: An Evaluation of Diagnostic Elements," *Dental Press Journal of Orthodontics* 15, no. 3 (May–June 2010): 134–57; Anita Bhavnani Rathod et al., "Extraction vs No Treatment: Long-Term Facial Profile Changes," *American Journal of Orthodontics and Dentofacial Orthopedics* 147, no. 5 (May 2015): 596–603; Abdol-Hamid Zafarmand and Mohamad-Mahdi Zafarmand, "Premolar Extraction in Orthodontics: Does It Have Any Effect on Patient's Facial Height?," *Journal of the International Society of Preventive & Community Dentistry* 5, no. 1 (Jan. 2015): 64–68.

124 **Brothers and sisters measured:** John Mew, *The Cause and Cure of Malocclusion* (John Mew Orthotropics), https://john meworthotropics.co.uk/the -cause-and-cure-of-malocclusion-book/; Vicki Cheeseman, interview with Kevin Boyd, "Understanding Modern Systemic Diseases through a Study of Anthropology," *Dentistry IQ,* June 27, 2012.

125 **Several other dentists:** More than two dozen scientific studies dating back to the 1930s are available at www.mr jamesnestor.com/breath.

125 **"quack," "scammer":** The half century of resistance from the orthodontics industry against John Mew, I'd learned, likely had less to do with Mew's data than with his take-no-prisoners approach to disseminating it. Even one of Mew's most ardent and vocal detractors, a British orthodontist named Roy Abrahams, admitted to me in an email exchange that it wasn't Mew's theories that were necessarily the problem, but that Mew had never proven his theories when given a chance and instead constantly "rubbish[es] traditional orthodontics and orthodontists to further his claims."

125 **famed evolutionary biologist:** Sandra Kahn and Paul R. Ehrlich, *Jaws: The Story of a Hidden Epidemic* (Stanford, CA: Stanford University Press, 2018).

127 **in his late 70s:** Mew told me that most of his foes use the castle as an example of how he'd profited from orthotropics. The total cost of the castle was about 300,000 pounds, he said, about a third of the cost of a dilapidated two-bedroom modern condo up the road.

127 **A 2006 peer-reviewed study:** G. Dave Singh et al., "Evaluation of the Posterior Airway Space Following Biobloc Therapy: Geometric Morphometrics," *Cranio: The Journal of Craniomandibular & Sleep Practice* 25, no. 2 (Apr. 2007): 84–89, https://facefocused.com/articles-and -lectures/bioblocs-impact-on-the-airway/.

128 **neck extended outward:** Assuming this open-mouthed posture throughout childhood can directly influence the growth and development of the jaws, airway, and even the alignment of the teeth. Joy L. Moeller et al., "Treating Patients with Mouth Breathing Habits: The Emerging Field of Orofacial Myofunctional Therapy," *Journal of the American Orthodontic Society* 12, no. 2 (Mar.–Apr. 2012): 10–12.

128 **"that's what we've become":** Modern humans could be the first *Homo* species to suffer from this malady. Even our Neanderthal cousins were not the knuckle-dragging, stooped-over beasts they'd been depicted as for the past hundred years. Their posture was upright, perhaps even better than our own. Martin Haeusler et al., "Morphology, Pathology, and the Vertebral Posture of the La Chapelle-aux-Saints Neandertal," *Proceedings of the National Academy of Sciences of the United States of America* 116, no. 11 (Mar. 2019): 4923–27.

128 **"cranial dystrophy":** M. Mew, "Craniofacial Dystrophy. A Possible Syndrome?," *British Dental Journal* 216, no. 10 (May 2014): 555–58.

129 **"a new health craze":** Elena Cresci, "Mewing Is the Fringe Orthodontic Technique Taking Over YouTube," *Vice,* Mar. 11, 2019, https://www.vice.com/en_us /article/d3medj/mewing-is-the-fringe -orthodontic-technique-taking-over -youtube.

129 **viewed a million times:** "Doing Mewing," YouTube, https://www.youtube.com/watch ?v=Hmf-pR7EryY.

130 **survival of the fittest:** Quentin Wheeler, Antonio G. Valdecasas, and Cristina Cânovas, "Evolution Doesn't Proceed in a Straight Line—So Why Draw It That Way?" *The Conversation,* Sept. 3, 2019, https:// theconversation.com/evolution-doesnt -proceed-in-a-straight-line-so-why-draw -it-that-way-109401/.

130 **Women will suffer:** "Anatomy & Physiology," Open Stax, Rice University, June 19, 2013, https://openstax.org/books/anatomy -and-physiology/pages/6-6-exercise -nutrition-hormones-and-bone-tissue.

130 **most apparent:** "Our Face Bones Change Shape As We Age," Live Science, May 30, 2013, https://www.livescience.com/35332 -face-bones-aging-110104.html.

131 **lead to airway obstruction:** Yagana Shah, "Why You Snore More As You Get Older and What You Can Do About It," *The Huffington Post,* June 7, 2015, https://

www.huffingtonpost.in/2015/07/06/how
-to-stop-snoring_n_7687906.html
?ri18n=true.

131 **power of the masseter:** "What Is the Strongest Muscle in the Human Body?," Everyday Mysteries: Fun Science Facts from the Library of Congress, https://www.loc.gov/rr/scitech/mysteries/muscles.html.

131 **more dense into our 70s:** Belfor wasn't the first researcher to discover this. In 1986, orthodontist Dr. Vincent G. Kokich, a professor in the Department of Orthodontics at the University of Washington and one of the world's experts in dentistry, postulated that adults "retain the capacity to regenerate and remodel bone at the craniofacial sutures." Liao, *Six-Foot Tiger*, 176–77.

132 **the more stem cells release:** We create stem cells throughout the body as well. The stem cells made in the sutures and jaws are often used for local maintenance in the mouth and face. Stem cells will ship off to whatever area needs them most. What they are attracted to are stress signals—in this case, the signals that come with vigorous chewing.

132 **two to four years of age:** "Weaning from the Breast," *Paediatrics & Child Health* 9, no. 4 (Apr. 2004): 249–53.

132 **lower incidence:** Bottle-feeding requires less "chewing" and sucking stress, and, as such stimulates less forward facial growth. For this reason, Kevin Boyd, the Chicago pediatric dentist, recommends cup feeding infants if breastfeeding isn't an option. James Sim Wallace, *The Cause and Prevention of Decay in Teeth* (London: J. & A. Churchill, 1902). Indrė Narbutyte et al., "Relationship Between Breastfeeding, Bottle-Feeding and Development of Malocclusion," *Stomatologija, Baltic Dental and Maxillofacial Journal* 15, no. 3 (2013): 67–72; Domenico Viggiano et al., "Breast Feeding, Bottle Feeding, and Non-Nutritive Sucking: Effects on Occlusion in Deciduous Dentition," *Archives of Disease in Childhood* 89, no. 12 (Jan. 2005): 1121–23; Bronwyn K. Brew et al., "Breastfeeding and Snoring: A Birth Cohort Study," *PLoS One* 9, no. 1 (Jan. 2014): e84956.

132 **stimulates the stress:** Every time I bite down while wearing the Homeblock I'll be eliciting cyclical intermittent light force in combination with the light springing pressure that will send a signal to the ligament around the roots of the teeth to encourage the body to, according to Belfor, "begin a cascade of events" which produces more bone cells. The process is called morphogenesis and it all sounded brutal. But Belfor assured me I won't even notice it happening because I'd only need to wear the retainer while I am sleeping.

133 **"An early soft diet":** Ben Miraglia, DDS, "2018 Oregon Dental Conference Course Handout," Oregon Dental Conference, Apr. 5, 2018, https://www.oregondental.org/docs/librariesprovider42/2018-odc-handouts/thursday---9122-miraglia.pdf?sfvrsn=2.

133 **ancient skulls measured:** Specifically, from 2.12 and 2.62 inches before the Industrial Age to 1.88 to 2.44 inches afterward. J. N. Starkey, "Etiology of Irregularities of the Teeth," *The Dental Surgeon* 4, no. 174 (Feb. 29, 1908): 105–6.

133 **"gradually becoming smaller":** J. Sim Wallace, "Heredity, with Special Reference to the Diminution in Size of the Human Jaw," digest of *Dental Record*, Dec. 1901, in *Dental Digest* 8, no. 2 (Feb. 1902): 135–40, https://tinyurl.com/r6szdz8.

134 **feeding a group of pigs:** Yucatan mini-pigs, that is. Russell L. Ciochon et al., "Dietary Consistency and Craniofacial Development Related to Masticatory Function in Minipigs," *Journal of Craniofacial Genetics and Developmental Biology* 17, no. 2 (Apr.–June 1997): 96–102.

134 **some form of malocclusion:** These round averages were summarized and verified by Dr. Robert Corruccini. Further overview context can be found in Mirigalia, "2018 Oregon Dental Conference Course Handout."

Chapter Eight: More, on Occasion

141 **twelve hundred men:** Micheal Clodfelter, *Warfare and Armed Conflicts: A Statistical Encyclopedia of Casualty and Other Figures, 1492–2015*, 4th ed. (Jefferson, NC: McFarland, 2017), 277.

141 **30 or more times:** J. M. Da Costa, "On Irritable Heart; a Clinical Study of a Form of Functional Cardiac Disorder and its Consequences," *American Journal of Medical Sciences*, n.s. 61, no. 121 (1871).

142 **The same symptoms:** Jeffrey A. Lieberman, "From 'Soldier's Heart' to 'Vietnam Syndrome': Psychiatry's 100-Year Quest to Understand PTSD," *The Star*, Mar. 7, 2015, https://www.thestar.com/news/insight/2015/03/07/solving-the-riddle-of-soldiers-heart-post-traumatic-stress-disorder-ptsd.html; Christopher Bergland. "Chronic Stress Can Damage Brain Structure and Connectivity," *Psychology Today*, Feb. 12, 2004.

142 **20 percent of soldiers:** "From Shell-Shock to PTSD, a Century of Invisible War Trauma," *PBS NewsHour*, Nov. 11, 2018, https://www.pbs.org/newshour/nation/from-shell-shock-to-ptsd-a-century-of-invisible-war-trauma; Caroline Alexander, "The Shock of War," *Smithsonian*, Sept. 2010, https://www.smithsonianmag.com/history/the-shock-of-war-55376701/#Mxod3dfdosgFt3cQ.99.

144 **breaths are so relaxing:** As well, the lower lungs contain from 60 to 80 percent of blood-saturated alveoli, for easier and more efficient gas exchange. *Body, Mind, and Sport*, 223.

144 **has an opposite role:** Phillip Low, "Overview of the Autonomic Nervous System," *Merck Manual*, consumer version, https://www.merckmanuals.com/home/brain,-spinal-cord,-and-nerve-disorders/autonomic-nervous-system-disorders/overview-of-the-autonomic-nervous-system.

145 **Heart rate increases:** "How Stress Can Boost Immune System," ScienceDaily, June 21, 2012; "Functions of the Autonomic Nervous System," Lumen, https://courses.lumenlearning.com/boundless-ap/chapter/functions-of-the-autonomic-nervous-system/.

145 **pupils dilate:** Joss Fong, "Eye-Opener: Why Do Pupils Dilate in Response to Emotional States?," *Scientific American*, Dec. 7, 2012, https://www.scientificamerican.com/article/eye-opener-why-do-pupils-dialate/.

145 **heightened sympathetic alert:** The sympathetic control center is located not in the brain but in the vertebral ganglia along the spine, while the parasympathetic system is located further up in the brain. This may not be a coincidence. Some researchers, such as Stephen Porges, suggest that the sympathetic system is a more primitive system while the parasympathetic system is more evolved.

145 **an hour or more:** "What Is Stress?," American Institute of Stress, https://www.stress.org/daily-life.

145 **man named Naropa:** "Tibetan Lama to Teach an Introduction to Tummo, the Yoga of Psychic Heat at HAC January 21," Healing Arts Center (St. Louis), Dec. 20, 2017, https://www.thehealingartscenter.com/hac-news/tibetan-lama-to-teach-an-introduction-to-tummo-the-yoga-of-psychic-heat-at-hac; "NAROPA," Garchen Buddhist Institute, July 14, 2015, https://garchen.net/naropa.

147 **wrote David-Néel:** Alexandra David-Néel, *My Journey to Lhasa* (1927; New York: Harper Perennial, 2005), 135.

148 **Professional surfers, mixed martial arts fighters:** Nan-Hie In, "Breathing Exercises, Ice Baths: How Wim Hof Method Helps Elite Athletes and Navy Seals," *South China Morning Post*, Mar. 25, 2019, https://www.scmp.com/lifestyle/health-wellness/article/3002901/wim-hof-method-how-ice-baths-and-breathing-techniques.

148 **His primary focus is the vagus:** Stephen W. Porges, *The Pocket Guide to the Polyvagal Theory: The Transformative Power of Feeling Safe*, Norton Series on Interpersonal Neurobiology (New York: W. W. Norton, 2017), 131, 140, 160, 173, 196, 242, 234.

149 **call it fainting:** Specifically, when the vagus nerve is stimulated, the heart rate slows and the blood vessels dilate, making it harder for blood to defeat gravity and be pumped to the brain. This temporary decrease in blood flow to the brain can cause the fainting episode.

150 **organs cut off from normal:** Steven Park, *Sleep Interrupted: A Physician Reveals the #1 Reason Why So Many of Us Are Sick and Tired* (New York: Jodev Press, 2008), Kindle locations 1443–46.

150 **lessen these symptoms:** "Vagus Nerve Stimulation," Mayo Clinic, https://www.mayoclinic.org/tests-procedures/vagus-nerve-stimulation/about/pac-20384565; Crystal T. Engineer et al., "Vagus Nerve Stimulation as a Potential Adjuvant to Behavioral Therapy for Autism and Other Neurodevelopmental Disorders," *Journal*

of Neurodevelopmental Disorders 9 (July 2017): 20.

150 **less invasive way:** There was also swinging. Rocking chairs and porch swings were very common in houses before the first half of the twentieth century. They may have been so popular because swinging shifts blood pressure, which allows messages to more easily travel back and forth along the vagus nerve. This is why so many autistic children (who often have poor vagal tone and feel constantly under threat) respond so well to swinging. Cold exposure, like splashing cold water on the face, also stimulates the vagus nerve, which sends messages to the heart to lower the heart rate. (Place your face in cold water and your heart rate will quickly drop.) Porges, *Pocket Guide to the Polyvagal Theory,* 211–12.

150 **speed up our heart:** Some very rare exceptions are demonstrated by yogis; these are discussed in the final chapter.

150 **when to breathe:** Roderik J. S. Gerritsen and Guido P. H. Band, "Breath of Life: The Respiratory Vagal Stimulation Model of Contemplative Activity," *Frontiers in Human Neuroscience* 12 (Oct. 2018): 397; Christopher Bergland, "Longer Exhalations Are an Easy Way to Hack Your Vagus Nerve," *Psychology Today,* May 9, 2019.

150 **Willing ourselves:** Moran Cerf, "Neuroscientists Have Identified How Exactly a Deep Breath Changes Your Mind," Quartzy, Nov. 19, 2017; Jose L. Herrero et al., "Breathing above the Brain Stem: Volitional Control and Attentional Modulation in Humans," *Journal of Neurophysiology* 119, no. 1 (Jan. 2018): 145–59.

150 **consciously access the autonomic:** The nervous system helps explain why breathing into a paper bag to control hyperventilation often doesn't work and can be very dangerous. Yes, capturing your exhaled breath will increase carbon dioxide levels, but it very often won't curb the sympathetic overload that could have triggered the panic attack to begin it. A paper bag might elicit more panic, and even deeper breathing. Further, not everyone who is suffering an attack in the respiratory system is suffering from hyperventilation. A study in *The Annals of Emergency Medicine* found that three patients believed to have been hyperventilating were given a paper bag to breathe into and died. These patients weren't suffering from a panic or asthma attack; they were suffering from a heart attack, and needed as much oxygen as they could get. Instead, they got lungs full of recycled carbon dioxide. Anahad O'Connor, "The Claim: If You're Hyperventilating, Breathe into a Paper Bag," *The New York Times,* May 13, 2008; Michael Callaham, "Hypoxic Hazards of Traditional Paper Bag Rebreathing in Hyperventilating Patients," *Annals of Emergency Medicine* 19, no. 6 (June 1989): 622–28.

150 **feeding and breeding:** Moran Cerf, "Neuroscientists Have Identified How Exactly a Deep Breath Changes Your Mind," Quartzy, Nov. 19, 2017; Jose L. Herrero, Simon Khuvis, Erin Yeagle, et al., "Breathing above the Brain Stem: Volitional Control and Attentional Modulation in Humans," *Journal of Neurophysiology* 119, no. 1 (Jan. 2018): 145–49.

150 **biologically impossible:** Matthijs Kox et al., "Voluntary Activation of the Sympathetic Nervous System and Attenuation of the Innate Immune Response in Humans," *Proceedings of the National Academy of Sciences of the United States of America* 111, no. 20 (May 2014): 7379–84.

151 **17 degrees Fahrenheit:** I wrote about Benson's work briefly in previous books and other writings, but in no instance did I explore what happens to the body and how, which is what I'm doing in this chapter.

151 **esteemed scientific journal:** Herbert Benson et al., "Body Temperature Changes during the Practice of g Tum-mo Yoga," *Nature* 295 (1982): 234–36. Decades later, not everyone was impressed by Benson's data. Maria Kozhevnikova, at the National University of Singapore, claimed there was "no evidence, however, indicating that temperatures are elevated beyond the normal range during g-tummo meditation." While she never dismissed the stunning effects of Tummo, Kozhevnikova wrote that the way in which the data was presented is misleading. With that, it should be noted that many Tummo practitioners told me that the exercise doesn't

so much get them hot; rather, *it keeps them from getting cold*, which has been clearly demonstrated both by the Buddhists and by Wim Hof and his crew. Either way, body heat is only a very small part of Tummo's transformative effects, as we'll soon learn. Maria Kozhevnikova et al., "Neurocognitive and Somatic Components of Temperature Increases during g-Tummo Meditation: Legend and Reality," *PLoS One* 8, no. 3 (2013): e58244.

151 **Arctic Circle shirtless:** "The Iceman—Wim Hof," Wim Hof Method, https://www .wimhofmethod.com/iceman-wim-hof.

152 **deepening his practice:** Erik Hedegaard, "Wim Hof Says He Holds the Key to a Healthy Life—But Will Anyone Listen?," *Rolling Stone*, Nov. 3, 2017.

152 **Andrew Huberman:** "Applications," Wim Hof Method, https://www.wimhof method.com/applications.

153 **two dozen healthy male volunteers:** Kox et al., "Voluntary Activation of the Sympathetic Nervous System."

153 **battery of immune cells:** "How Stress Can Boost Immune System," Science-Daily, June 21, 2012, https://www.science daily.com/releases/2012/06 /120621223525.htm.

153 **self-produced opioids:** Joshua Rapp Learn, "Science Explains How the Iceman Resists Extreme Cold," Smithsonian.com, May 22, 2018.

154 **50 million people:** The National Institutes of Health estimates that up to 23.5 million Americans suffer from autoimmune disease. The American Autoimmune Related Disease Association says this number is a gross underestimate because the NIH only lists 24 diseases associated with autoimmune disorder; however, there are several dozen other diseases not listed which have a clear "autoimmune basis." You can read the sobering statistics at https://www .aarda.org/.

155 **Hashimoto's disease:** New research shows that narcolepsy is also an autoimmune disease, and perhaps even asthma. That children with asthma have a 41 percent increased risk of getting type 1 diabetes is likely no coincidence. Alberto Tedeschi and Riccardo Asero, "Asthma and Autoimmunity: A Complex but Intriguing Relation," *Expert Review of Clinical Immunology* 4, no. 6 (Nov. 2008): 767–76; Natasja Wulff Pedersen et al., "CD8+ T Cells from Patients with Narcolepsy and Healthy Controls Recognize Hypocretin Neuron-Specific Antigens," *Nature Communications* 10, no. 1 (Feb. 2019): 837.

155 **I'd heard dozens:** Before trying Tummo, Matt had been diagnosed with psoriatic arthritis and had C-reactive protein (CRP), which contributes to the inflammation and soreness in his disease, of over 20, about seven times the normal level. After three months of practicing Tummo breathing with exposure to cold, Matt's CRP levels were 0.4. All of the soreness in his joints, the stiffness, the flaky red skin, and the fatigue were gone. Another Matt, from Devon in England, was diagnosed with lichen planopilaris, an inflammatory disease that mainly affects the scalp and results in scaling and permanent patchy hair loss. Matt was put on a prescription for hydroxychloroquine, a medicine invented in 1955 to treat malaria that suppresses the immune response. Common side effects of hydroxychloroquine include cramps, diarrhea, headaches, and worse. Within a week Matt was having trouble breathing and was coughing up blood. His doctor told him to push through. Matt got sicker. He learned Tummo breathing and followed Hof's protocol, practicing the Wim Hof Method every day. Wim Hof, YouTube, Jan. 3, 2018, https://www.you tube.com/watch?v=f4tIou2LnOk; "Wim Hof—Reversing Autoimmune Diseases | Paddison Program," YouTube, June 26, 2016, https://www.youtube.com/watch?v= lZO9uyJIP44; "In 8 Months I Was Completely Symptom-Free," Wim Hof Method Experience, Wim Hof, YouTube, Aug. 23, 2019, https://www.youtube.com/watch?v= 1nO v4aNiWys.

155 **reducing inflammatory markers:** In 2014, Hof took a group of 26 random people, aged 29 to 65, up Mount Kilimanjaro. Many in the group suffered from asthma, rheumatism, Crohn's, and other autoimmune dysfunctions. He taught them his version of Tummo breathing, exposed them to periodic bouts of extreme cold, then hiked 19,300 feet to the top of Africa's tallest mountain. Oxygen levels at the top are half of what they are at sea level. The

success rate of experienced climbers is about 50 percent. Twenty-four of Hof's students, including those with autoimmune disorders, made it to the summit in 48 hours. Half the group ascended bare-chested, wearing nothing but shorts in temperatures that dip to minus 4 degrees Fahrenheit. None experienced hypothermia or altitude sickness, and none used supplemental oxygen. Ted Thornhill, "Hardy Climbers Defy Experts to Reach Kilimanjaro Summit Wearing Just Their Shorts and without Succumbing to Hypothermia," *Daily Mail*, Feb. 17, 2014; "Kilimanjaro Success Rate—How Many People Reach the Summit," Kilimanjaro, https://www.climbkilimanjaroguide.com/kilimanjaro-success-rate. An older estimate put the number at 41 percent; the current estimate is probably closer to 60 percent. I've split the difference.

157 **David-Néel used:** It's worth noting that David-Néel eventually became a national hero in France, an idol to the Beat writers, and had a tea and a tram station named after her, both of which are still in use today.

157 **Maurice Daubard:** "Maurice Daubard—Le Yogi des Extrêmes [The Yogi of the Extremes]," http://www.mauricedaubard.com/biographie.htm; "France: Moulins: Yogi Maurice Daubard Demonstration," AP Archive, YouTube, July 21, 2015, https://www.youtube.com/watch?time_continue=104&v=bEZVlgcddZg.

158 **Stanislav Grof:** This interview and my experience with Holotropic Breathwork happened several years before the Stanford experiment, and just a year or so after that jarring experience with Sudarshan Kriya that sent me on the path to deeper research.

159 **It was November 1956:** Grof had told me this event took place in 1954; however, other sources claim it occurred in 1956. "The Tim Ferriss Show—Stan Grof, Lessons from ~4,500 LSD Sessions and Beyond," Podcast Notes, Nov. 24, 2018, https://podcastnotes.org/2018/11/24/grof/.

159 **The experience would guide:** "Stan Grof," Grof: Know Thyself, http://www.stanislavgrof.com.

159 **By 1968, the U.S. government:** Mo Costandi, "A Brief History of Psychedelic Psychiatry," *The Guardian*, Sept. 2, 2014, https://www.theguardian.com/science/neurophilosophy/2014/sep/02/psychedelic-psychiatry.

160 **Eyerman led more than 11,000:** James Eyerman, "A Clinical Report of Holotropic Breathwork in 11,000 Psychiatric Inpatients in a Community Hospital Setting," *MAPS Bulletin*, Spring 2013, http://www.maps.org/news-letters/v23n1/v23n1_24-27.pdf.

161 **"figure out why":** Eyerman continued: "When you think about it, Western industrial civilization is the only group in entire human history that doesn't hold nonordinary states of consciousness in high esteem, [that] doesn't appreciate and want to understand them," he told me. "Instead, we pathologize them, we numb them with tranquilizers. This works as a band-aid works, as a temporary fix, but it does not address the core problem and just leads to more psychological problems later on."

161 **smaller studies followed:** Sarah W. Holmes et al., "Holotropic Breathwork: An Experiential Approach to Psychotherapy," *Psychotherapy: Theory, Research, Practice, Training* 33, no. 1 (Spring 1996): 114–20; Tanja Miller and Laila Nielsen, "Measure of Significance of Holotropic Breathwork in the Development of Self-Awareness," *Journal of Alternative and Complementary Medicine* 21, no. 12 (Dec. 2015): 796–803; Stanislav Grof et al., "Special Issue: Holotropic Breathwork and Other Hyperventilation Procedures," *Journal of Transpersonal Research* 6, no. 1 (2014); Joseph P. Rhinewine and Oliver Joseph Williams, "Holotropic Breathwork: The Potential Role of a Prolonged, Voluntary Hyperventilation Procedure as an Adjunct to Psychotherapy," *Journal of Alternative and Complementary Medicine* 13, no. 7 (Oct. 2007): 771–76.

163 **having less oxygen in the brain:** Specifically, all that huffing depleted our bloodstream of carbon dioxide, and thus, cut off blood flow the brain needs to function properly. Stanislav Grof and Christina Grof, *Holotropic Breathwork: A New Approach to Self-Exploration and Therapy*, SUNY Series in Transpersonal

and Humanistic Psychology (Albany, NY: Excelsior, 2010), 161, 163; Stanislav Grof, *Psychology of the Future: Lessons from Modern Consciousness Research* (Albany, NY: SUNY Press, 2000); Stanislav Grof, "Holotropic Breathwork: New Approach to Psychotherapy and Self-Exploration," http://www.stanislavgrof.com/resources /Holotropic-Breathwork;-New-Perspec tives-in-Psychotherapy-and-Self-Explora tion.pdf.

163 **During rest, about 750 milliliters:** "Cerebral Blood Flow and Metabolism," Neurosurg.cam.ac.uk, http://www.neurosurg .cam.ac.uk/files/2017/09/2-Cerebral -blood-flow.pdf.

163 **Blood flow can increase:** Jordan S. Querido and A. William Sheel, "Regulation of Cerebral Blood Flow during Exercise," *Sports Medicine* 37, no. 9 (2007): 765–82.

163 **decrease by 40 percent:** On average, cerebral blood flow will reduce about 2 percent for every 1mmHg decrease of carbon dioxide in the blood ($PaCO_2$). During one heavy breathing exercise recording at a laboratory at the University of California, San Francisco, my $PaCO_2$ clocked in at 22 mmHg, which is about 20 below normal. During that time my brain was receiving about 40 percent less blood flow than normal. "Hyperventilation," OpenAnesthesia, https://www.openanes thesia.org/elevated_icp_hyperventilation.

163 **The areas most affected:** An interesting summary, including several scientific studies, is available on this webpage: http://www.anesthesiaweb.org/hyperven tilation.php.

163 **signals throughout the body:** "Rhythm of Breathing Affects Memory and Fear," *Neuroscience News*, Dec. 7, 2016, https:// neurosciencenews.com/memory -fear-breathing-5699/.

Chapter Nine: Hold It

165 **A couple years later:** Details of Kling's research and the following account of S. M. were drawn from Justin S. Feinstein et al., "A Tale of Survival from the World of Patient S. M.," in *Living without an Amygdala*, ed. David G. Amaral and Ralph Adolphs (New York: Guilford Press, 2016), 1–38. Other details were pulled

from Kling's articles, including Arthur Kling et al., "Amygdalectomy in the Free-Ranging Vervet (*Cercopithecus aethiops*)," *Journal of Psychiatric Research* 7, no. 3 (Feb. 1970): 191–99.

166 **alarm circuit of fear:** "The Amygdala, the Body's Alarm Circuit," Cold Spring Harbor Laboratory DNA Learning Center, https://dnalc.cshl.edu/view/822-The-Amygdala-the-Body-s-Alarm-Circuit .html.

169 **cluster of neurons:** We have two kinds of chemoreceptors in our respiratory system: peripheral and central. Peripheral chemoreceptors, in the carotid artery and aorta, are for the most part responsible for detecting changes in the amount of oxygen in the blood as it leaves the heart. Central chemoreceptors, located in the brain stem, detect very minute changes in the levels of carbon dioxide within arterial blood via the pH of the cerebrospinal fluid. "Chemoreceptors," TeachMe Physiology, https:// teachmephysiology.com/respiratory -system/regulation/chemoreceptors.

170 **flex and shift with changing environments:** People with injuries in the area of the brain stem that holds the central chemoreceptors lose the ability to gauge and react to carbon dioxide levels in the bloodstream. With no autonomic trigger to alert them that carbon dioxide is building, each breath they take requires a conscious and concerted effort. They'll suffocate in their sleep without a respirator because their bodies won't know when to breathe. The condition is called Ondine's disease, and gets its name from a water sprite in a European folk tale. Ondine told her husband, Hans, that she was "the breath in [his] lungs," and warned him that if he ever cheated on her he would lose his ability to unconsciously breathe. Hans cheated and suffered from Ondine's curse. "A single moment of inattention and I forget to breathe," Hans said before he died. Iman Feiz-Erfan et al., "Ondine's Curse," *Barrow Quarterly* 15, no. 2 (1999), https://www.barrowneuro.org/education /grand-rounds-publications-and-media /barrow-quarterly/volume-15-no-2-1999 /ondines-curse/.

170 **altitudes 800 feet below and 16,000 feet:** Twelve thousand years ago, ancient

Peruvians inhabited enclaves 12,000 feet above sea level. The current highest inhabited city is La Rinconada, Peru, at an elevation of 16,728 feet above sea level. Tia Ghose, "Oldest High-Altitude Human Settlement Discovered in Andes," Live Science, Oct. 23, 2014, https://www.livescience.com/48419-high-altitude-setllement-peru.html;

170 **some elite mountain climbers:** According to some reports, athletes like freedivers tend to have about the same carbon dioxide tolerance as people who are not acclimated to taking repeated, very long breath holds. The hypothesis is that such top-tier athletes have much larger lungs and may also be able to slow their metabolism down to such a level that they consume less oxygen and produce less carbon dioxide, allowing them to hold their breath for longer without feeling anxious. But this doesn't explain why people with chronic anxieties and other fear-based disorders almost always have very limited breath-holding ability, regardless of their lung size or how much they inhaled or exhaled before the test. Some interesting (if not limited) context can be found on the Deeper Blue freediving forum: https://forums.deeperblue.com/threads/freediving-leading-to-sleep-apnea.82096/. Colette Harris, "What It Takes to Climb Everest with No Oxygen," *Outside*, June 8, 2017, https://www.outsideonline.com/2191596/how-train-climb-everest-no-oxygen.

171 **Eighteen percent of Americans:** Jamie Ducharme, "A Lot of Americans Are More Anxious Than They Were Last Year, a New Poll Says," *Time*, May 8, 2018, https://time.com/5269371/americans-anxiety-poll/.

171 **offered this advice:** *The Primordial Breath: An Ancient Chinese Way of Prolonging Life through Breath Control*, vol. 1, trans. Jane Huang and Michael Wurmbrand (Original Books, 1987), 13.

172 **"terribly damaging":** See a detailed explanation of the damages caused by oxidative stress and nitric oxide synthase by Dr. Scott Simonetti at www.mrjamesnestor.com/breath.

172 **continuous partial attention:** Megan Rose Dickey, "Freaky: Your Breathing Patterns Change When You Read Email," *Business Insider*, Dec. 5, 2012, https://www.businessinsider.com/email-apnea-how-email-change-breathing-2012-12?IR=T; "Email Apnea," Schott's Vocab, *The New York Times*, Sept. 23, 2009, https://schott.blogs.nytimes.com/2009/09/23/email-apnea/; Linda Stone, "Just Breathe: Building the Case for Email Apnea," *The Huffington Post*, https://www.huffpost.com/entry/just-breathe-building-the_b_85651; Susan M. Pollak, "Breathing Meditations for the Workplace," *Psychology Today*, Nov. 6, 2014, https://www.psychologytoday.com/us/blog/the-art-now/201411/email-apnea.

173 **out of our control:** Dozens of studies are accessible through the United States National Library of Medicine at the National Institutes of Health website PubMed. Here's a few that were helpful to me: Andrzej Ostrowski et al., "The Role of Training in the Development of Adaptive Mechanisms in Freedivers," *Journal of Human Kinetics* 32, no. 1 (May 2012): 197–210; Apar Avinash Saoji et al., "Additional Practice of Yoga Breathing With Intermittent Breath Holding Enhances Psychological Functions in Yoga Practitioners: A Randomized Controlled Trial," *Explore: The Journal of Science and Healing* 14, no. 5 (Sept. 2018): 379–84; Saoji et al., "Immediate Effects of Yoga Breathing with Intermittent Breath Holding on Response Inhibition among Healthy Volunteers," *International Journal of Yoga* 11, no. 2 (May–Aug. 2018): 99–104.

174 **to war wounds:** Serena Gianfaldoni et al., "History of the Baths and Thermal Medicine," *Macedonian Journal of Medical Sciences* 5, no. 4 (July 2017): 566–68.

174 **"cured almost with certainty":** George Henry Brandt, *Royat (les Bains) in Auvergne, Its Mineral Waters and Climate* (London: H. K. Lewis, 1880), 12, 18; Peter M. Prendergast and Melvin A. Shiffman, eds., *Aesthetic Medicine: Art and Techniques* (Berlin and Heidelberg: Springer, 2011); William and Robert Chambers, *Chambers's Edinburgh Journal*, n.s. 1, no. 46 (Nov. 16, 1844): 316; Isaac Burney Yeo, *The Therapeutics of Mineral Springs and Climates* (London: Cassell, 1904), 760.

174 **"The study" . . . a British doctor:** After Brandt returned to Britain and raved about Royat, another doctor and fellow at

the Royal College of Surgeons left to visit Royat to confirm Brandt's findings, and reported them "quite in accordance with my own experience and observations." George Henry Brandt, *Royat (les Bains) in Auvergne: Its Mineral Waters and Climate* (London: H. K. Lewis, 1880), 12, 18.

175 **scientific research disappeared:** According to Dr. Lewis S. Coleman, a California anesthesiologist and medical researcher, the backlash against carbon dioxide probably had less to do with the facts and more to do with private interests. Carbon dioxide was an inexpensive by-product of oil processing, whereas other clinical treatments were expensive and required real expertise to administer. Lewis S. Coleman, "Four Forgotten Giants of Anesthesia History," *Journal of Anesthesia and Surgery* 3, no. 1 (2016): 68–84.

175 **skin disorders:** See dozens of studies on the benefits of carbon dioxide bathing at mrjamesnestor.com/breath.

176 **Joseph Wolpe . . . Donald Klein:** In the late 1950s, Wolpe was looking for alternative treatments for free-floating anxiety, a form of stress for which there is no specific cause, which today affects about 10 million Americans. He was floored by how quickly and effectively carbon dioxide worked. Between two and five inhalations of a 50/50 mixture of carbon dioxide and oxygen, Wolpe found, was enough to lower the baseline level of anxiety in his patients from 60 (debilitating) to zero. No other treatment came close. "It will be hoped that the recently awakened interest in carbon dioxide will lead to active research," wrote Wolpe in 1987. But the same year Wolpe published his carbon dioxide call to arms, the Food and Drug Administration approved the first SSRI drug, fluoxetine, which would become better known by its trade names Prozac, Sarafem, and Adofen. A decade after Wolpe's study was published, Donald F. Klein, a Columbia University psychiatrist, found what he thought was the mechanism that triggered panic, anxiety, and related disorders. It was a "physiologic misinterpretation by a suffocation monitor [that] misfires an evolved suffocation alarm system," wrote Klein in his paper "False Suffocation Alarms, Spontaneous

Panics, and Related Conditions." And that false suffocation was coming from chemoreceptors that had grown to become too sensitive to fluctuations in carbon dioxide. Fear, at its core, could be as much a physical problem as a mental one. Joseph Wolpe, "Carbon Dioxide Inhalation Treatments of Neurotic Anxiety: An Overview," *Journal of Nervous and Mental Disease* 175, no. 3 (Mar. 1987): 129–33; Donald F. Klein, "False Suffocation Alarms, Spontaneous Panics, and Related Conditions," *Archives of General Psychiatry* 50, no. 4 (Apr. 1993): 206–17.

176 **half of us will suffer:** This is Feinstein's estimate. Hard numbers are difficult to pin down because so many people with anxiety disorders suffer from depression, and vice versa. For instance, an estimated 18 percent of the population suffers from anxiety disorders; some 8 percent suffer from major depressive disorder, and millions more with milder systems; one quarter suffer from a diagnosable mental disorder; and one half of all Americans are expected to suffer from some mental illness throughout their lives. "Half of US Adults Due for Mental Illness, Study Says," Live Science, Sept. 1, 2011, https://www.livescience.com/15876-mental-illness-strikes-adults.html; "Facts & Statistics," Anxiety and Depression Association of America, https://adaa.org/about-adaa/press-room/facts-statistics

176 **13 percent:** Further, depression, anxiety, and panic are all closely related, each rooted in the same misinterpretation of fear. A third of patients currently on SSRIs suffer from other forms of anxiety and many will be treated with different drugs for those conditions. Laura A. Pratt et al., "Antidepressant Use Among Persons Aged 12 and Over: United States, 2011–2014," NCHS Data Brief no. 283 (Aug. 2017): 1–8.

176 **described as "weak":** These findings, as you might imagine, were controversial. You can read more about the ongoing debate of this study in Fredrik Hieronymus et al., "Influence of Baseline Severity on the Effects of SSRIs in Depression: An Item-Based, Patient-Level Post-Hoc Analysis," *The Lancet*, July 11, 2019, https://www.thelancet.com/journals/lanpsy/article/PIIS2215-0366(19)30383-9/fulltext;

Fredrik Hieronymus, "How Do We Determine Whether Antidepressants Are Useful or Not? Authors' Reply," *The Lancet*, Nov. 2019, https://www.thelancet.com/journals/lanpsy/article/PIIS2215-0366(19)30383-9/fulltext; Henry Bodkin, "Most Common Antidepressant Barely Helps Improve Depressive Symptoms, 'Shocking' Trial Finds," *The Telegraph* (UK), Sept. 19, 2019, https://www.telegraph.co.uk/science/2019/09/19/common-antidepressant-barely-helps-improve-depression-symptoms.

177 **exposure therapy:** An overview of treatments and efficacy is available here: Johanna S. Kaplan and David F. Tolin, "Exposure Therapy for Anxiety Disorders," *Psychiatric Times*, Sept. 6, 2011, https://www.psychiatrictimes.com/anxiety/exposure-therapy-anxiety-disorders.

177 **anorexia or panic:** Some 40 percent of panic disorder patients suffer from depression, and 70 percent have some other mental health conditions. All of these conditions, Feinstein says, are rooted in fear. Paul M. Lehrer, "Emotionally Triggered Asthma: A Review of Research Literature and Some Hypotheses for Self-Regulation Therapies," *Applied Psychophysiology and Biofeedback* 22, no. 1 (Mar. 1998): 13–41.

177 **holding their breath:** Panic sufferers visit the doctor five times more often than other patients, and are six times more likely to be hospitalized for psychiatric disorders. Thirty-seven percent of them will seek some treatment, usually drugs, behavior therapy, or both. But none of these therapies directly address what could be contributing to this condition: chronic poor breathing habits. That 60 percent of people with chronic obstructive pulmonary disease also have anxiety or depressive disorders is not a coincidence. These patients are very often breathing too much, too fast, panicking in anticipation of not being able to take another breath. "Proper Breathing Brings Better Health," *Scientific American*, Jan. 15, 2019, https://www.scientificamerican.com/article/proper-breathing-brings-better-health/.

177 **hypersensitized to carbon dioxide:** Eva Henje Blom et al., "Adolescent Girls with Emotional Disorders Have a Lower End-Tidal CO_2 and Increased Respiratory Rate Compared with Healthy Controls," *Psychophysiology* 51, no. 5 (May 2014): 412–18; Alicia E. Meuret et al., "Hypoventilation Therapy Alleviates Panic by Repeated Induction of Dyspnea," *Biological Psychiatry CNNI (Cognitive Neuroscience and Neuroimaging)* 3, no. 6 (June 2018): 539–45; Daniel S. Pine et al., "Differential Carbon Dioxide Sensitivity in Childhood Anxiety Disorders and Nonill Comparison Group," *Archives of General Psychiatry* 57, no. 10 (Oct. 2000): 960–67.

177 **Alicia Meuret:** "Out-of-the-Blue Panic Attacks Aren't without Warning: Data Show Subtle Changes before Patients' [*sic*] Aware of Attack," Southern Methodist University Research, https://blog.smu.edu/research/2011/07/26/out-of-the-blue-panic-attacks-arent-without-warning/; Stephanie Pappas, "To Stave Off Panic, Don't Take a Deep Breath," Live Science, Dec. 26, 2017, https://www.livescience.com/9204-stave-panic-deep-breath.html.

177 **capnometers, which recorded:** "New Breathing Therapy Reduces Panic and Anxiety by Reversing Hyperventilation," ScienceDaily, Dec. 22, 2010, https://www.sciencedaily.com/releases/2010/12/101220200010.htm; Pappas, "To Stave Off Panic."

178 **soundless room:** Floatation, as Feinstein has found through five years of clinical research, was particularly effective in treating anxiety, anorexia, and other fear-based neuroses. "The Feinstein Laboratory," Laureate Institute for Brain Research, http://www.laureateinstitute.org/current-events/feinstein-laboratory-publishes-float-study-in-plos-one.

180 **"super endurance":** See Buteyko's chart of optimum (and dangerously low) carbon dioxide levels at https://images.app.goo.gl/DGjT3bL8PMDQYmqL7.

180 **reports from pulmonauts:** Recently, carbon dioxide therapy has made a bit of a comeback, not just with Olsson and his crew of DIY pulmonauts. It's now being used, again, to treat hearing loss, epilepsy, and various cancers. The U.S. health care provider Aetna offers carbon dioxide therapy as an experimental treatment for patients. "Carbogen Inhalation Therapy," Aetna, http://www.aetna.com/cpb/medical/data/400_499/0428.html.

181 **chemoreceptors are normally:** Chemoreceptors designed to analyze the tiniest fluctuations in carbon dioxide, a fraction of one percent.

Chapter Ten: Fast, Slow, and Not at All

186 **heavy dose of stress hormones:** Even an hour after the Tummo practice ended. Think of the lungs as a solar panel; the larger the panel, the more cells there are to soak up sunlight, the more available energy. Wim Hof's heavy breathing can increase available space for gas exchange by about 40 percent—a tremendous amount. With this bonus space, Hof, for instance, was able to consume double the normal amount of oxygen 40 minutes after he finished the exercises. Isabelle Hof, *The Wim Hof Method Explained* (Wim Hof Method, 2015, updated 2016), 8, https://explore.wimhofmethod.com/wp-content/uploads/ebook-the-wim-hof-method-explained-EN.pdf.

186 **snow for hours and not:** Joshua Rapp Learn, "Science Explains How the Iceman Resists Extreme Cold," Smithsonian.com, May 22, 2018, https://www.smithsonianmag.com/science-nature/science-explains-how-iceman-resists-extreme-cold-180969134/#WUf1Swaj7zYCkVDv.99.

186 **breathe slow and less:** Herbert Benson et al., "Body Temperature Changes during the Practice of g Tum-mo Yoga," *Nature* 295 (1982): 234–36; William J. Cromie, "Meditation Changes Temperatures," *The Harvard Gazette*, Apr. 18, 2002.

187 **they certainly are not:** I queried this conundrum to Dr. Paul Davenport, a renowned physiologist and Distinguished Professor at the University of Florida. He replied within a few hours. "Interesting problem," he wrote in an email. "My answer will be appropriately, academically vague :) Bottom line, the effect of voluntary hyperventilation depends on multiple factors including regional blood distribution, degree of blood gas changes, reduced buffering capacity of cerebrospinal fluid (CSF), changes in cardiac output, pH balance compensation, time and factors yet unknown. (Is that ambiguous enough?) Research on the physiological response of

blood and CSF to voluntary hyperventilation is relatively straight forward. However, the cognitive responses to the physiological changes is much more ambiguous and complex." At the end of the email, he told me he was working on a detailed analysis of the problem, which would take some time to put together. As of this writing, he was still writing it. I will post it on my website: mrjamesnestor.com/breath. In the interim, you can peruse a few studies here: I. A. Bubeev, "The Mechanism of Breathing under the Conditions of Prolonged Voluntary Hyperventilation," *Aerospace and Environmental Medicine* 33, no. 2 (1999): 22–26; J. S. Querido and A. W. Sheel, "Regulation of Cerebral Blood Flow during Exercise," *Sports Medicine* 37, no. 9 (Oct. 2007), 765–82.

187 **mechanism behind these techniques:** Iuriy A. Bubeev and I. B. Ushakov, "The Mechanism of Breathing under the Conditions of Prolonged Voluntary Hyperventilation," *Aerospace and Environmental Medicine* 33, no. 2 (1999): 22–26; Seymour S. Kety and Carl F. Schmidt, "The Effects of Altered Arterial Tensions of Carbon Dioxide and Oxygen on Cerebral Blood Flow and Cerebral Oxygen Consumption of Normal Young Men," *Journal of Clinical Investigation* 27, no. 4 (1948): 484–92; Querido and Sheel, "Regulation of Cerebral Blood Flow during Exercise"; Shinji Naganawa et al., "Regional Differences of fMR Signal Changes Induced by Hyperventilation: Comparison between SE-EPI and GE-EPI at 3-T," *Journal of Magnetic Resonance Imaging* 15, no. 1 (Jan. 2002): 23–30; S. Posse et al., "Regional Dynamic Signal Changes during Controlled Hyperventilation Assessed with Blood Oxygen Level-Dependent Functional MR Imaging," *American Journal of Neuroradiology* 18, no. 9 (Oct. 1997): 1763–70.

188 **same time in India and China:** More specifically, written references to prana appeared in India about 3,000 years ago, and in China during the Ying and Zhou periods about 2,500 years ago.

188 **prana power lines:** Ancient Indians believed the body contained from 72,000 to 350,000 channels. How they might have counted them, nobody knows.

188 **never observed prana:** Sat Bir Singh Khalsa et al., *Principles and Practice of Yoga in Health Care* (Edinburgh: Handspring, 2016).

188 **confirmed that it exists:** There was, however, some very weird, and fascinating, government-supported research into the possibilities of moving this "vital energy." Check out this gem of a study from 1986 that somehow seeped through the cracks of the CIA website: Lu Zuyin et al., "Physical Effects of Qi on Liquid Crystal," CIA, https://www.cia.gov/library/reading room/docs/CIA-RDP96-00792R00020 0160001-8.pdf.

188 **a group of physicists:** Justin O'Brien (Swami Jaidev Bharati), *Walking with a Himalayan Master: An American's Odyssey* (St. Paul, MN: Yes International, 1998, 2005), 58, 241; Pandit Rajmani Tigunait, *At the Eleventh Hour: The Biography of Swami Rama* (Honesdale, PA: Himalayan Institute Press, 2004); "Swami Rama, Researcher/Scientist," Swami Rama Society, http://www.swamiramasociety.org/project/swami-rama-researcherscientist/.

189 **By the age of three:** "Swami Rama, Himalayan Master, Part 1," YouTube, https://www.youtube.com/watch?v=S1sZN bRH2N8.

189 **a small, pictureless office:** "Swami Rama at the Menninger Clinic, Topeka, Kansas," Kansas Historical Society, https://www.kshs.org/index.php?url=km/items/view/226459.

189 **Veterans Administration hospital:** Dr. Daniel Ferguson, chief of the medical hygiene clinic of the Veterans Administration Hospital in Minnesota, had shown a few months earlier that Swami Rama had the ability to make his pulse "disappear" for minutes at a time. Erik Peper et al., eds., *Mind/Body Integration: Essential Readings in Biofeedback* (New York: Plenum Press, 1979), 135.

190 **for 17 seconds:** Actual recorded time was 17 seconds, though Rama had entered into this heart-fluttering zone several seconds before the technicians were prepared. This detail was taken from Justin O'Brien's *The Wellness Tree: The Six-Step Program for Creating Optimal Wellness* (Yes International, 2000).

190 **The results of the experiment:** Gay Luce and Erik Peper, "Mind over Body, Mind over Mind," *The New York Times*, Sept. 12, 1971.

190 **Within 15 minutes:** Marilynn Wei and James E. Groves, *The Harvard Medical School Guide to Yoga* (New York: Hachette, 2017); Jon Shirota, "Meditation: A State of Sleepless Sleep," June 1973, http://hihtin dia.org/wordpress/wp-content/uploads/2012/10/swamiramaprobe1973.pdf.

191 **television talk shows:** "Swami Rama: Voluntary Control over Involuntary States," YouTube, Jan. 22, 2017, 1:17, https://www.youtube.com/watch?v=yv_D3ATDvVE.

191 **French cardiologist:** Mathias Gardet, "Thérèse Brosse (1902–1991)," https://re penf.hypotheses.org/795; "Biofeedback Research and Yoga," Yoga and Consciousness Studies, http://www.yogapsychology.org/art_biofeedback.html; Brian Luke Seaward, *Managing Stress: Principles and Strategies for Health and Well-Being* (Burlington, MA: Jones & Bartlett Learning, 2012); M. A. Wenger and B. K. Bagchi, "Studies of Autonomic Functions in Practitioners of Yoga in India," *Behavioral Science* 6, no. 4 (Oct. 1961): 312–23.

191 **with the goal of reaching:** "Swami Rama Talks: 2:1 Breathing Digital Method," Swami Rama. YouTube, May 23, 2019, https://www.youtube.com/watch?v=PYVrB36FrQw; "Swami Rama Talks: OM Kriya pt. 1," Swami Rama. YouTube, May 28, 2019, https://www.youtube.com/watch?v=ygvnWEnvWCQ.

191 **Rama obviously had learned:** Rama, apparently, wasn't all peacefulness and light. In 1994 a female student who'd attended the Himalayan Institute charged that Rama had initiated sexual abuse when she was 19 and he was in his late 60s. Four years later, after Rama's death, a jury awarded the woman almost two million dollars in damages. Management at the Himalayan Institute contend the trial was unfair, as Rama was not even present to offer his side of the story. Nonetheless, the incident stained Rama's legacy at home and abroad. William J. Broad, "Yoga and Sex Scandals: No Surprise Here," *The New York Times*, Feb. 27, 2012.

192 **Albert Szent-Gyorgyi:** Biographical information is summarized from the

following sources: Robyn Stoller, "The Full Story of Dr. Albert Szent-Györgyi," National Foundation for Cancer Research, Dec. 9, 2017, https://www.nfcr.org/blog/full-story-of-dr-albert-szent-gyorgyi/; Albert Szent-Györgyi, "Biographical Overview," National Library of Medicine, https://profiles.nlm.nih.gov/spotlight/wg/feature/biographical; Robert A. Kyle and Marc A. Shampo, "Albert Szent-Györgyi—Nobel Laureate," *Mayo Clinic Proceedings* 75, no. 7 (July 2000): 722; "Albert Szent-Györgyi: Scurvy: Scourge of the Sea," Science History Institute, https://www.sciencehistory.org/historical-profile/albert-szent-gyorgyi.

192 **"All living organisms":** Albert Szent-Györgyi, "Muscle Research," *Scientific American* 180 (June 1949): 22–25.

193 **the more *alive* it is:** According to researchers at the University of Arizona, in Tucson, what separated animals with small brains from those with large and rapidly evolving brains was their capacity for endurance exercise. The higher the capacity, the larger the brain. What fueled this capacity, and these brains, were larger lungs capable of better respiratory efficiency. This helps explain why mammals have larger brains than non-mammals, and why human, whale, and dolphin brains kept growing so rapidly over millions of years as reptilian brains didn't. Oxygen equals energy equals evolution. Our ability to breathe large and full breaths, in some ways, helped make us human. David A. Raichlen and Adam D. Gordon, "Relationship between Exercise Capacity and Brain Size in Mammals," *PLoS One* 6, no. 6 (June 2011): e20601; "Functional Design of the Respiratory System," medicine.mcgill.ca, https://www.medicine.mcgill.ca/physio/resp-web/TEXT1.htm; Alexis Blue, "Brain Evolved to Need Exercise," Neuroscience News, June 26, 2017, https://neurosciencenews.com/evolution-brain-exercise-6982/.

193 **millions of years:** Bettina E. Schirrmeister et al., "Evolution of Multicellularity Coincided with Increased Diversification of Cyanobacteria and the Great Oxidation Event," *PNAS* 110, no. 5 (Jan. 2013): 1791–96.

193 **"the living state":** Albert Szent-Györgyi, "The Living State and Cancer," *Physiological Chemistry and Physics*, Dec. 1980.

193 **"simple but subtle":** Szent-Györgyi attributes this phrase to personal communication with P. Ehrenfest, an Austrian-Dutch theoretical physicist.

193 **begin to break down:** G. E. W. Wolstenholme et al., eds., *Submolecular Biology and Cancer* (Hoboken, NJ: John Wiley & Sons, 2008): 143.

193 **environments of low oxygen:** J. Cui et al., "Hypoxia and Miscoupling between Reduced Energy Efficiency and Signaling to Cell Proliferation Drive Cancer to Grow Increasingly Faster," *Journal of Molecular Cell Biology*, 2012; Alexander Greenhough et al., "Cancer Cell Adaptation to Hypoxia Involves a HIF-GPRC5A-YAP Axis," *EMBO Molecular Medicine*, 2018.

194 **"In every culture and in every medical tradition":** This quotation has been attributed to Szent-Györgyi's lecture "Electronic Biology and Cancer," which he presented at the Marine Biological Laboratory, Woods Hole, Massachusetts, July, 1972.

195 **filled with copies:** "Master DeRose," enacademic.com, https://enacademic.com/dic.nsf/enwiki/11708766.

195 **now Afghanistan, Pakistan:** The Indus Valley descriptions and details are taken from the following: "Indus River Valley Civilizations," Khan Academy, https://www.khanacademy.org/humanities/world-history/world-history-beginnings/ancient-india/a/the-indus-river-valley-civilizations; Saifullah Khan, "Sanitation and Wastewater Technologies in Harappa /Indus Valley Civilization (ca. 2600–1900 bce)," https://canvas.brown.edu/files/61957992/download?download_frd=1.

195 **largest geographically:** To put this in perspective, 300,000 square miles is equivalent to all East Coast states from Florida to New York. Craig A. Lockard, *Societies, Networks, and Transitions: A Global History* (Stamford, CT: Cengage Learning, 2008).

196 **A seal engraving:** Yan Y. Dhyansky, "The Indus Valley Origin of a Yoga Practice," *Artibus Asiae* 48, nos. 1–2 (1987), pp. 89–108.

196 **birthplace of yoga:** A thorough description of the history, epistemology, and evolution of Samkhya and the earliest yoga can be found in this excellent academic

paper at the *Internet Encyclopedia of Philosophy*, https://www.iep.utm.edu/yoga/.

196 **of Nazi lore:** The word *Aryan* comes from the Sanskrit *ērān*, which was the basis for the modern country name of Iran. The term never had anything to do with white supremacy until the Nazis appropriated it some four thousand years later.

196 **language of Sanskrit:** Steve Farmer et al., "The Collapse of the Indus-Script Thesis: The Myth of a Literate Harappan Civilization," *Electronic Journal of Vedic Studies* 11, no. 2 (Jan. 2014): 19–57, http://laurasi anacademy.com/ejvs/ejvs1102/ejvs1102 article.pdf.

196 **Chandogya Upanishads:** From a philosophy called Samkhya. Samkhya was based on reason and proof. The noun root of Samkhya means "number"; the verb root means "to know." "Either you know or you don't," DeRose told me. "Spirituality had nothing to do with it!" The foundation of Samkhya was secular, based on empirical studies, not opinions. He told me that there was no mention of any praying hands or standing yoga postures in the earliest Upanishads because these exercises were never part of the practice. The earliest yoga was a technology developed to influence and control prana. It was a science of meditation and breathing. Possibly the earliest reference to pranayama (the ancient Indian art of breath control) is listed in hymn 1.5.23 of the Brihadaranyaka Upanishad, which was first documented around 700 BCE. "One should indeed breathe in (arise), but one should also breathe out (without setting) while saying, 'Let *not* the misery that is dying reach me.' When one would practice that (breathing), one should rather desire to thoroughly realize that (immortality). It is rather through *that* (realization) that he wins a union with this divinity (breath), that is a sharing of worlds." *The Brihadaranyaka Upanishad*, book 1, trans. John Wells, Darshana Press, http://darshana press.com/Brihadaranyaka%20Upani shad%20Book%201.pdf.

196 **India, China, and beyond:** By the sixth century BCE, Siddhartha Gautama, the son of an Indus Valley warrior king and queen, found his way to beneath a ficus

tree in northeast India. He sat down and started practicing these ancient breathing and meditation techniques. Gautama became enlightened, and took off to teach the wonders of breathing, meditation, and enlightenment throughout the East. Siddhartha would later become known as the Buddha, founder of the Buddhist faith.

196 **By around 500 BCE:** Michele Marie Desmarais, *Changing Minds: Mind, Consciousness and Identity in Patanjali's Yoga-sutra and Cognitive Neuroscience* (Delhi: Motilal Banarsidass, 2008).

196 **extending exhalations:** The actual passage is much more vague. According to DeRose, it translates to something along these lines: "The fourth type of pranayama transcends inhalation and exhalation." Interpretations of the Yoga Sutras vary broadly; the interpretation I listed, by Swami Jnaneshvara, I found the most elucidating and accessible. More here: http://swamij.com/yoga-sutras-24953.htm, http://www.swamij.com/yoga-sutras-24953.htm#2.51.

197 **the instructors would yell:** Mestre DeRose, *Quando É Preciso Ser Forte: Autobiografia* (Portuguese edition) (São Paulo: Egrégora, 2015).

197 **only in the twentieth century:** After Patanjali, yoga was further compressed and rewritten. The Bhagavad Gita described it as more of a mystical and metaphysical practice, a spiritual tool to be used to bring self-realization and enlightenment. The Hatha tradition of yoga, which was formally developed in the 1400s, used the ancient techniques to honor the Lord Shiva and converted the seated asanas into 15 poses, many of them in a standing position. "Contesting Yoga's Past: A Brief History of Āsana in Pre-modern India," Center for the Study of World Religions, Oct. 14, 2015, https://cswr.hds.harvard .edu/news/2015/10/14/contesting -yoga's-past-brief-history-āsana-pre -modern-india.

198 **An estimated two billion people:** "Two Billion People Practice Yoga 'Because It Works,'" UN News, June 21, 2016, https:// news.un.org/en/audio/2016/06/614172; Alice G. Walton, "How Yoga Is Spreading in

the U.S.," *Forbes*, https://www.forbes.com/sites/alicegwalton/2016/03/15/how-yoga-is-spreading-in-the-u-s/#3809c047449f/.

198 **But what have we lost:** In his book *Pranayama* (I received a prepublication copy), DeRose details 58 breathing techniques whose roots go back thousands of years to the origins of Samkhya. A few of these techniques are offered at the end of this book.

198 **Krishna to Jesus Christ:** "The Most Ancient and Secretive Form of Yoga Practiced by Jesus Christ: Kriya Yoga," Evolve+ Ascend, http://www.evolveandascend.com/2016/05/24/ancient-secretive-form-yoga-practiced-jesus-christ-kriya-yoga; "The Kriya Yoga Path of Meditation," Self-Realization Fellowship, https://www.yogananda-srf.org/The_Kriya_Yoga_Path_of_Meditation.aspx.

199 **tens of millions of people:** "Research on Sudarshan Kriya Yoga," Art of Living, https://www.artofliving.org/us-en/research-sudarshan-kriya.

199 **Art of Living:** I can't describe how to do Sudarshan Kriya because there are no written instructions. Shankar is the only one who conducts these sessions, and does so through a crackly old recording like the one I heard so many years ago. Anyone who wants to try Sudarshan Kriya will need to head out to an Art of Living outpost or scour the Internet for a bootleg. I've done both.

200 **or any other breathing practice:** This is one reason why randomly hyperventilating or practicing nontraditional breathing techniques can be so damaging and dangerous.

Epilogue: A Last Gasp

205 **"More than sixty years":** Albert Szent-Györgyi, "The Living State and Cancer," in G. E. W. Wolstenholme et al., eds., *Submolecular Biology and Cancer* (Hoboken, NJ: John Wiley & Sons, 2008), 17.

205 **the top killers:** "The Top 10 Causes of Death," World Health Organization, May 24, 2018, https://www.who.int/newsroom/fact-sheets/detail/the-top-10-causes-of-death; "Leading Causes of Death," Centers for Disease Control and Prevention, https://www.cdc.gov/nchs/fastats/leading-causes-of-death.htm.

205 **Genes can be turned off:** Danielle Simmons, "Epigenetic Influences and Disease," Nature Education, https://www.nature.com/scitable/topicpage/epigenetic-influences-and-disease-895/.

206 **30 pounds of air:** "Each day about 30 pounds of air participates in this tidal flow, compared with less than 4 pounds of food and 5 pounds of water." Dr. John R. Goldsmith, "How Air Pollution Has Its Effect on Health (2)—Air Pollution and Lung Function Changes," *Proceedings: National Conference on Air Pollution U.S. Department of Health, Education, and Welfare* (Washington, DC: United States Government Printing Office, 1959), 215.

206 **"If I had to":** Andrew Weil, *Breathing: The Master Key to Self Healing*, Sounds True, 1999.

208 **bacterial infection:** I still had remnants of bacterial infestation in my nose, but it was almost nonexistent. The results: "A 2+ *Corynebacterium propinquum*: Rare number Gram Positive Cocci; rare to small number of Gram Positive Rods; No polymorphonuclear cells."

208 **overcoming congestion:** "Nasal congestion" is a blanket term for any obstruction in the nose caused by the mucus membrane or sinuses. Rhinologists have two names for the most common forms of congestion. The first is rhinitis, which means an inflammation of the mucus membrane in the nose. The second is sinusitis, which means the inflammation in the cavities surrounding the nasal passages. The numbers on exactly how many people are suffering from one or the other, or both, are hard to pin down, as some people may be suffering from both chronically, and others may have bouts of either rhinitis or sinusitis separately. Seasonal trends affect these statistics as well. Rhinitis is considered the fifth most common disease, affecting up to 30 to 40 percent of the population. Sinusitis affects up to 30 percent of the population. These are rough numbers; several neurologists I talked to claimed the percentages were far higher. R. M. Rosenfeld et al., "Clinical Practice Guideline (Update): Adult Sinusitis Executive Summary," *Otolaryngology—Head and Neck Surgery* 152, no. 4 (April 2015): 598–609.

209 **"The difference in breathing":** Carl Stough and Reece Stough, *Dr. Breath: The Story of Breathing Coordination* (New York: William Morrow, 1970), 29.

210 **rougher, rawer, and heartier foods:** Charles Matthews, "Just Eat What Your Great-Grandma Ate," *San Francisco Chronicle*, Dec. 30, 2007, https://michael pollan.com/reviews/just-eat-what-your -great-grandma-ate/.

INDEX